SYSTEMS ENGINEERING FUNDAMENTALS

United States Government
US Army

Consider Amazon.com for all your Professional Books.
Fast shipping, high quality, and low prices.

TABLE OF CONTENTS

PREFACE iv

PART 1. INTRODUCTION
    Chapter 1.   Introduction to Systems Engineering Management 3
    Chapter 2.   Systems Engineering Management in DoD Acquisition 11

PART 2. THE SYSTEMS ENGINEERING PROCESS
    Chapter 3.   Systems Engineering Process Overview 31
    Chapter 4.   Requirements Analysis 35
    Chapter 5.   Functional Analysis and Allocation 45
    Chapter 6.   Design Synthesis 57
    Chapter 7.   Verification 65
    Chapter 8.   Systems Engineering Process Outputs 73

PART 3. SYSTEM ANALYSIS AND CONTROL
    Chapter 9.   Work Breakdown Structure 85
    Chapter 10. Configuration Management 91
    Chapter 11. Technical Reviews and Audits 99
    Chapter 12. Trade Studies 111
    Chapter 13. Modeling and Simulation 117
    Chapter 14. Metrics 125
    Chapter 15. Risk Management 133

PART 4. PLANNING, ORGANIZING, AND MANAGING
    Chapter 16. Systems Engineering Planning 147
    Chapter 17. Product Improvement Strategies 157
    Chapter 18. Organizing and Integrating System Development 171
    Chapter 19. Contractual Considerations 185
    Chapter 20. Management Considerations and Summary 201

GLOSSARY 209

PREFACE

This book provides a basic, conceptual-level description of engineering management disciplines that relate to the development and life cycle management of a system. For the non-engineer it provides an overview of how a system is developed. For the engineer and project manager it provides a basic framework for planning and assessing system development.

Information in the book is from various sources, but a good portion is taken from lecture material developed for the two Systems Planning, Research, Development, and Engineering courses offered by the Defense Acquisition University.

The book is divided into four parts: *Introduction; Systems Engineering Process; Systems Analysis and Control;* and *Planning, Organizing, and Managing.* The first part introduces the basic concepts that govern the systems engineering process and how those concepts fit the Department of Defense acquisition process. Chapter 1 establishes the basic concept and introduces terms that will be used throughout the book. The second chapter goes through a typical acquisition life cycle showing how systems engineering supports acquisition decision making.

The second part introduces the systems engineering problem-solving process, and discusses in basic terms some traditional techniques used in the process. An overview is given, and then the process of requirements analysis, functional analysis and allocation, design synthesis, and verification is explained in some detail. This part ends with a discussion of the documentation developed as the finished output of the systems engineering process.

Part three discusses analysis and control tools that provide balance to the process. Key activities (such as risk management, configuration management, and trade studies) that support and run parallel to the system engineering process are identified and explained.

Part four discusses issues integral to the conduct of a systems engineering effort, from planning to consideration of broader management issues.

In some chapters supplementary sections provide related material that shows common techniques or policy-driven processes. These expand the basic conceptual discussion, but give the student a clearer picture of what systems engineering means in a real acquisition environment.

PART 1

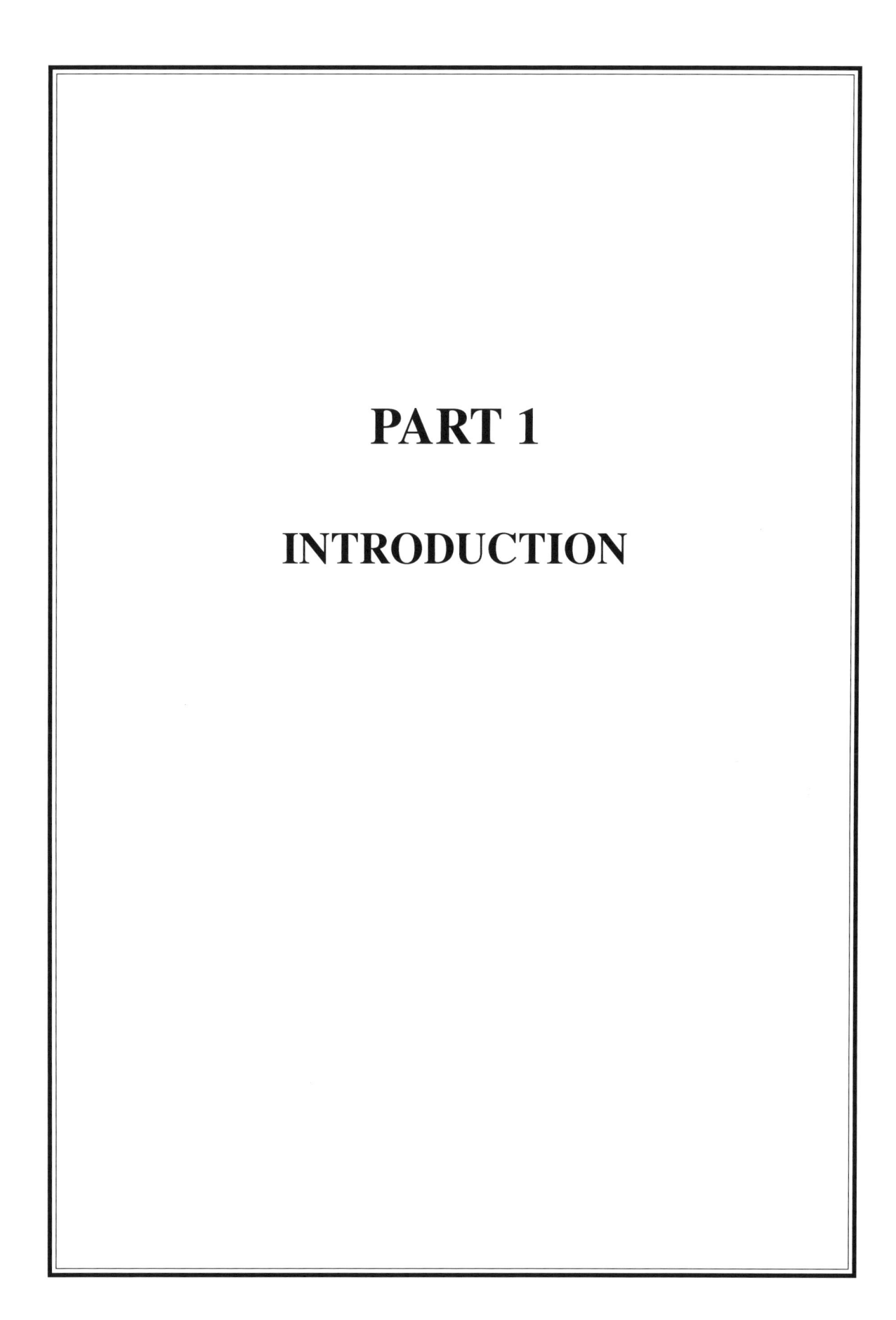

INTRODUCTION

CHAPTER 1

INTRODUCTION TO SYSTEMS ENGINEERING MANAGEMENT

1.1 PURPOSE

The overall organization of this text is described in the Preface. This chapter establishes some of the basic premises that are expanded throughout the book. Basic terms explained in this chapter are the foundation for following definitions. Key systems engineering ideas and viewpoints are presented, starting with a definition of a system.

1.2 DEFINITIONS

A System Is ...

Simply stated, a system is an integrated composite of people, products, and processes that provide a capability to satisfy a stated need or objective.

Systems Engineering Is...

Systems engineering consists of two significant disciplines: the technical knowledge domain in which the systems engineer operates, and systems engineering management. This book focuses on the process of systems engineering management.

Three commonly used definitions of systems engineering are provided by the best known technical standards that apply to this subject. They all have a common theme:

- A logical sequence of activities and decisions that transforms an operational need into a description of system performance parameters and a preferred system configuration. (MIL-STD-499A, *Engineering Management*, 1 May 1974. Now cancelled.)

- An interdisciplinary approach that encompasses the entire technical effort, and evolves into and verifies an integrated and life cycle balanced set of system people, products, and process solutions that satisfy customer needs. (EIA Standard IS-632, *Systems Engineering*, December 1994.)

- An interdisciplinary, collaborative approach that derives, evolves, and verifies a life-cycle balanced system solution which satisfies customer expectations and meets public acceptability. (IEEE P1220, *Standard for Application and Management of the Systems Engineering Process*, [Final Draft], 26 September 1994.)

In summary, systems engineering is an interdisciplinary engineering management process that evolves and verifies an integrated, life-cycle balanced set of system solutions that satisfy customer needs.

Systems Engineering Management Is...

As illustrated by Figure 1-1, systems engineering management is accomplished by integrating three major activities:

- Development phasing that controls the design process and provides baselines that coordinate design efforts,

- A systems engineering process that provides a structure for solving design problems and

3

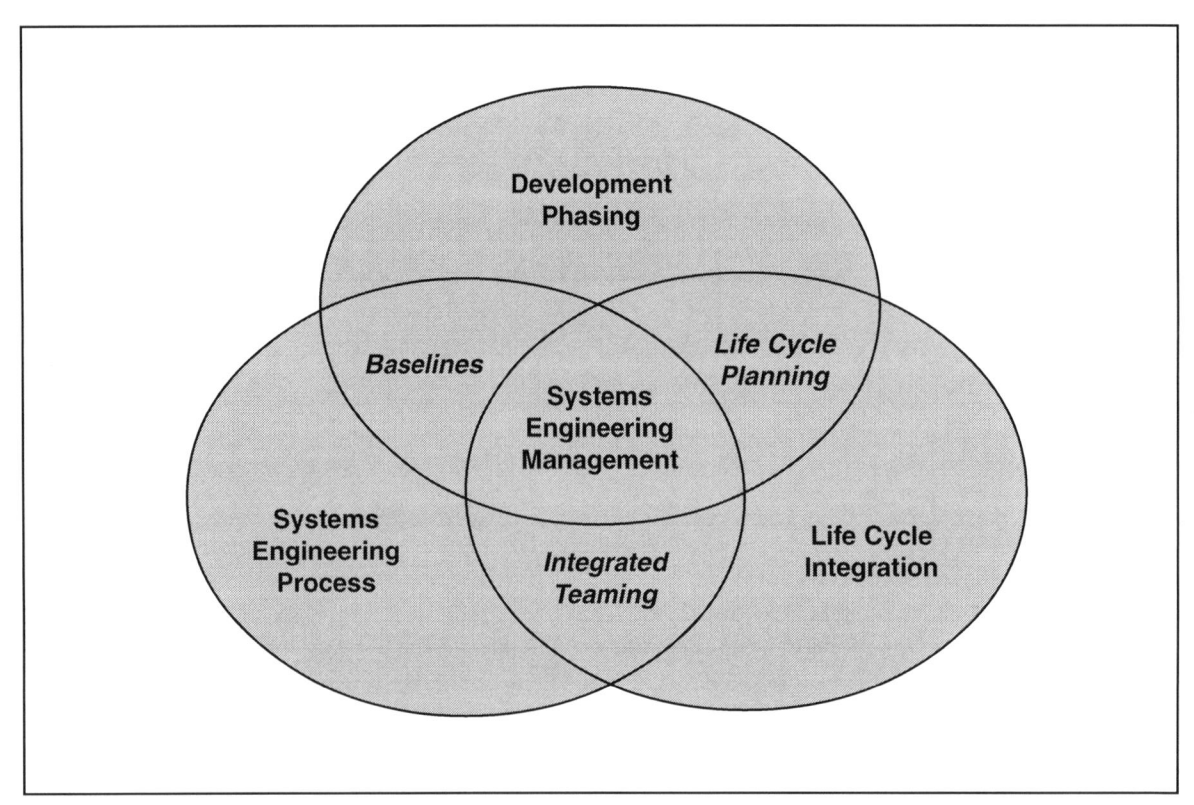

Figure 1-1. Three Activities of Systems Engineering Management

tracking requirements flow through the design effort, and

- Life cycle integration that involves customers in the design process and ensures that the system developed is viable throughout its life.

Each one of these activities is necessary to achieve proper management of a development effort. Phasing has two major purposes: it controls the design effort and is the major connection between the technical management effort and the overall acquisition effort. It controls the design effort by developing design baselines that govern each level of development. It interfaces with acquisition management by providing key events in the development process, where design viability can be assessed. The viability of the baselines developed is a major input for acquisition management Milestone (MS) decisions. As a result, the timing and coordination between technical development phasing and the acquisition schedule is critical to maintain a healthy acquisition program.

The systems engineering process is the heart of systems engineering management. Its purpose is to provide a structured but flexible process that transforms requirements into specifications, architectures, and configuration baselines. The discipline of this process provides the control and traceability to develop solutions that meet customer needs. The systems engineering process may be repeated one or more times during any phase of the development process.

Life cycle integration is necessary to ensure that the design solution is viable throughout the life of the system. It includes the planning associated with product and process development, as well as the integration of multiple functional concerns into the design and engineering process. In this manner, product cycle-times can be reduced, and the need for redesign and rework substantially reduced.

1.3 DEVELOPMENT PHASING

Development usually progresses through distinct levels or stages:

- Concept level, which produces a system concept description (usually described in a concept study);

- System level, which produces a system description in performance requirement terms; and

- Subsystem/Component level, which produces first a set of subsystem and component product performance descriptions, then a set of corresponding detailed descriptions of the products' characteristics, essential for their production.

The systems engineering process is applied to each level of system development, one level at a time, to produce these descriptions commonly called configuration baselines. This results in a series of configuration baselines, one at each development level. These baselines become more detailed with each level.

In the Department of Defense (DoD) the configuration baselines are called the functional baseline for the system-level description, the allocated baseline for the subsystem/component performance descriptions, and the product baseline for the subsystem/component detail descriptions. Figure 1-2 shows the basic relationships between the baselines. The triangles represent baseline control decision points, and are usually referred to as technical reviews or audits.

Levels of Development Considerations

Significant development at any given level in the system hierarchy should not occur until the configuration baselines at the higher levels are considered complete, stable, and controlled. Reviews and audits are used to ensure that the baselines are ready for the next level of development. As will be shown in the next chapter, this review and audit process also provides the necessary assessment of system maturity, which supports the DoD Milestone decision process.

1.4 THE SYSTEMS ENGINEERING PROCESS

The systems engineering process is a top-down comprehensive, iterative and recursive problem

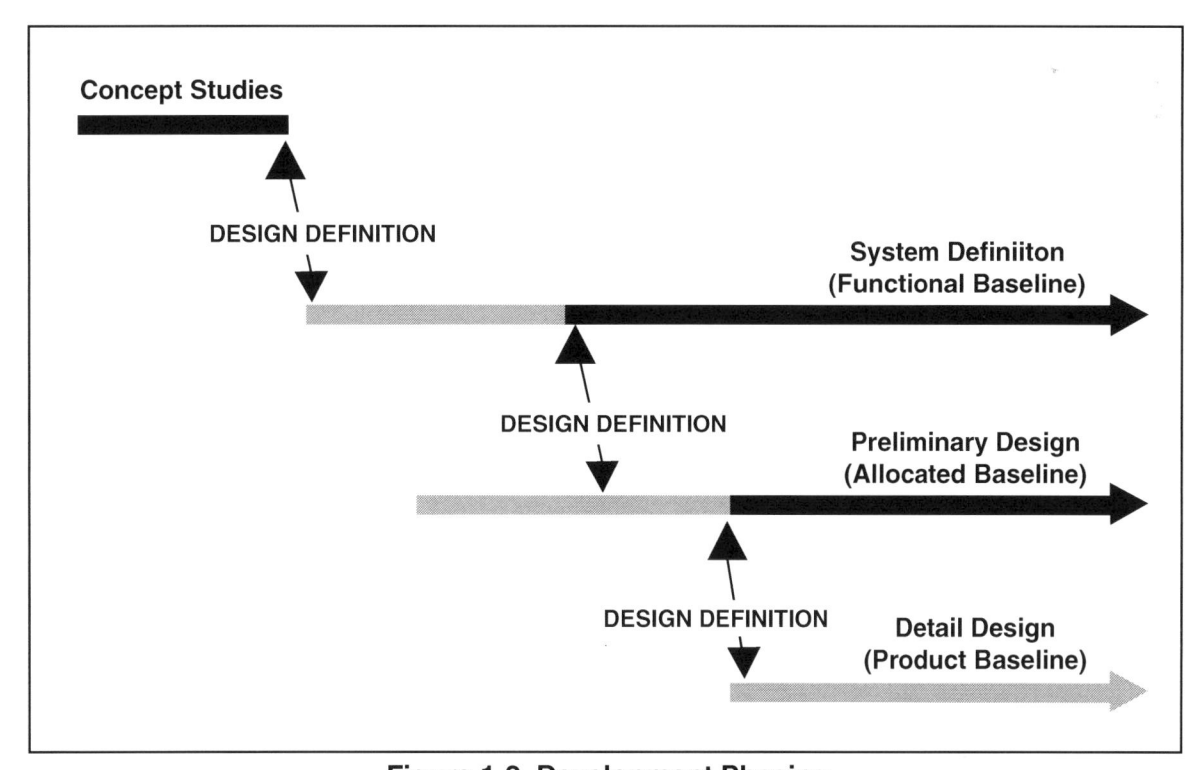

Figure 1-2. Development Phasing

solving process, applied sequentially through all stages of development, that is used to:

- Transform needs and requirements into a set of system product and process descriptions (adding value and more detail with each level of development),

- Generate information for decision makers, and

- Provide input for the next level of development.

As illustrated by Figure 1-3, the fundamental systems engineering activities are Requirements Analysis, Functional Analysis and Allocation, and Design Synthesis—all balanced by techniques and tools collectively called System Analysis and Control. Systems engineering controls are used to track decisions and requirements, maintain technical baselines, manage interfaces, manage risks, track cost and schedule, track technical performance, verify requirements are met, and review/audit the progress.

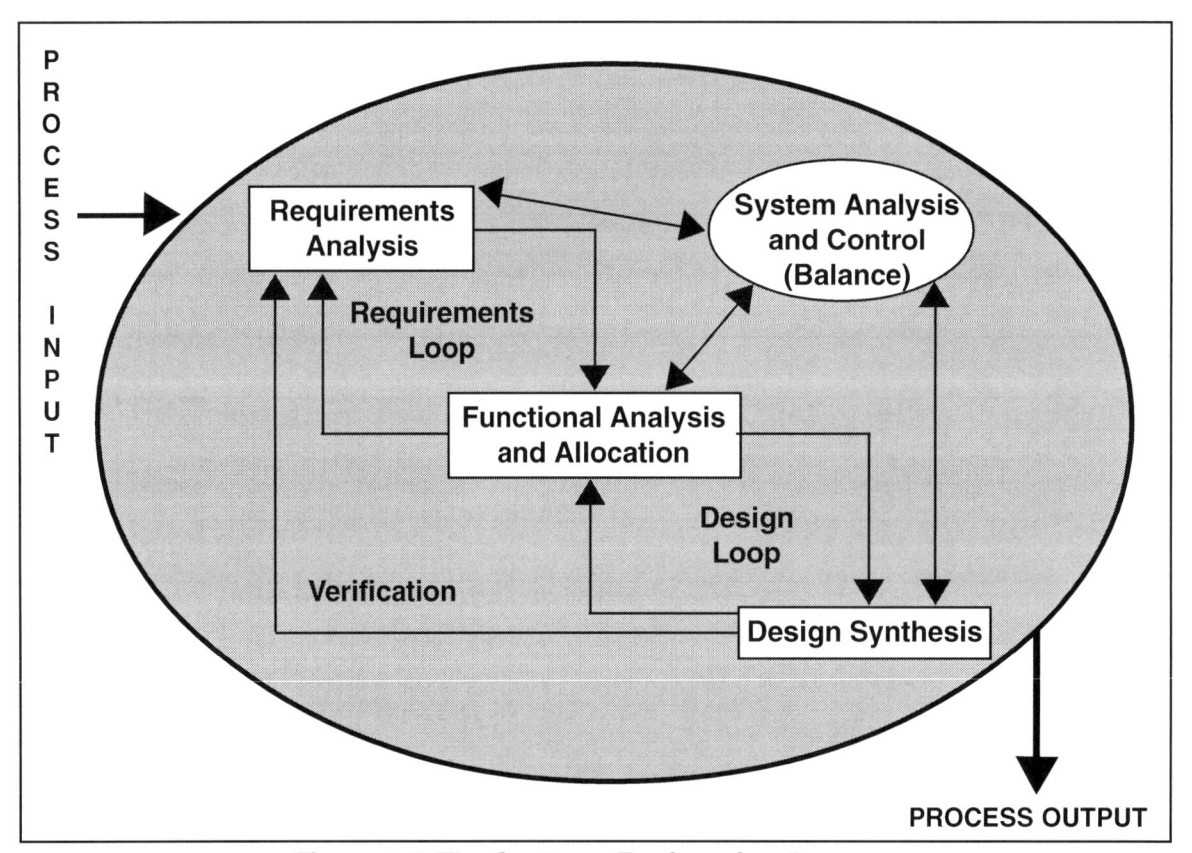

During the systems engineering process architectures are generated to better describe and understand the system. The word "architecture" is used in various contexts in the general field of engineering. It is used as a general description of how the subsystems join together to form the system. It can also be a detailed description of an aspect of a system: for example, the Operational, System, and Technical Architectures used in Command, Control, Communications, Computers, Intelligence, Surveillance, and Reconnaissance (C4ISR), and software intensive developments. However, Systems Engineering Management as developed in DoD recognizes three universally usable architectures that describe important aspects of the system: functional, physical, and system architectures. This book will focus on these architectures as necessary components of the systems engineering process.

The *Functional Architecture* identifies and structures the allocated functional and performance requirements. The *Physical Architecture* depicts the

Figure 1-3. The Systems Engineering Process

system product by showing how it is broken down into subsystems and components. The *System Architecture* identifies all the products (including enabling products) that are necessary to support the system and, by implication, the processes necessary for development, production/construction, deployment, operations, support, disposal, training, and verification.

Life Cycle Integration

Life cycle integration is achieved through integrated development—that is, concurrent consideration of all life cycle needs during the development process. DoD policy requires integrated development, called Integrated Product and Product Development (IPPD) in DoD, to be practiced at all levels in the acquisition chain of command as will be explained in the chapter on IPPD. Concurrent consideration of all life cycle needs can be greatly enhanced through the use of interdisciplinary teams. These teams are often referred to as Integrated Product Teams (IPTs).

The objective of an Integrated Product Team is to:

- Produce a design solution that satisfies initially defined requirements, and

- Communicate that design solution clearly, effectively, and in a timely manner.

Multi-functional, integrated teams:

- Place balanced emphasis on product and process development, and

- Require early involvement of all disciplines appropriate to the team task.

Design-level IPT members are chosen to meet the team objectives and generally have distinctive competence in:

- Technical management (systems engineering),

- Life cycle functional areas (eight primary functions),

- Technical specialty areas, such as safety, risk management, quality, etc., or

- When appropriate, business areas such as finance, cost/budget analysis, and contracting.

Life Cycle Functions

Life cycle functions are the characteristic actions associated with the system life cycle. As illustrated by Figure 1-4, they are development, production and construction, deployment (fielding), operation, support, disposal, training, and verification. These activities cover the "cradle to grave" life cycle process and are associated with major functional groups that provide essential support to the life cycle process. These key life cycle functions are commonly referred to as the eight primary functions of systems engineering.

The customers of the systems engineer perform the life-cycle functions. The system user's needs are emphasized because their needs generate the requirement for the system, but it must be remembered that all of the life-cycle functional areas generate requirements for the systems engineering process once the user has established the basic need. ***Those that perform the primary functions also provide life-cycle representation in design-level integrated teams.***

Primary Function Definitions

Development includes the activities required to evolve the system from customer needs to product or process solutions.

Manufacturing/Production/Construction includes the fabrication of engineering test models and "brass boards," low rate initial production, full-rate production of systems and end items, or the construction of large or unique systems or subsystems.

Deployment (Fielding) includes the activities necessary to initially deliver, transport, receive, process, assemble, install, checkout, train, operate, house, store, or field the system to achieve full operational capability.

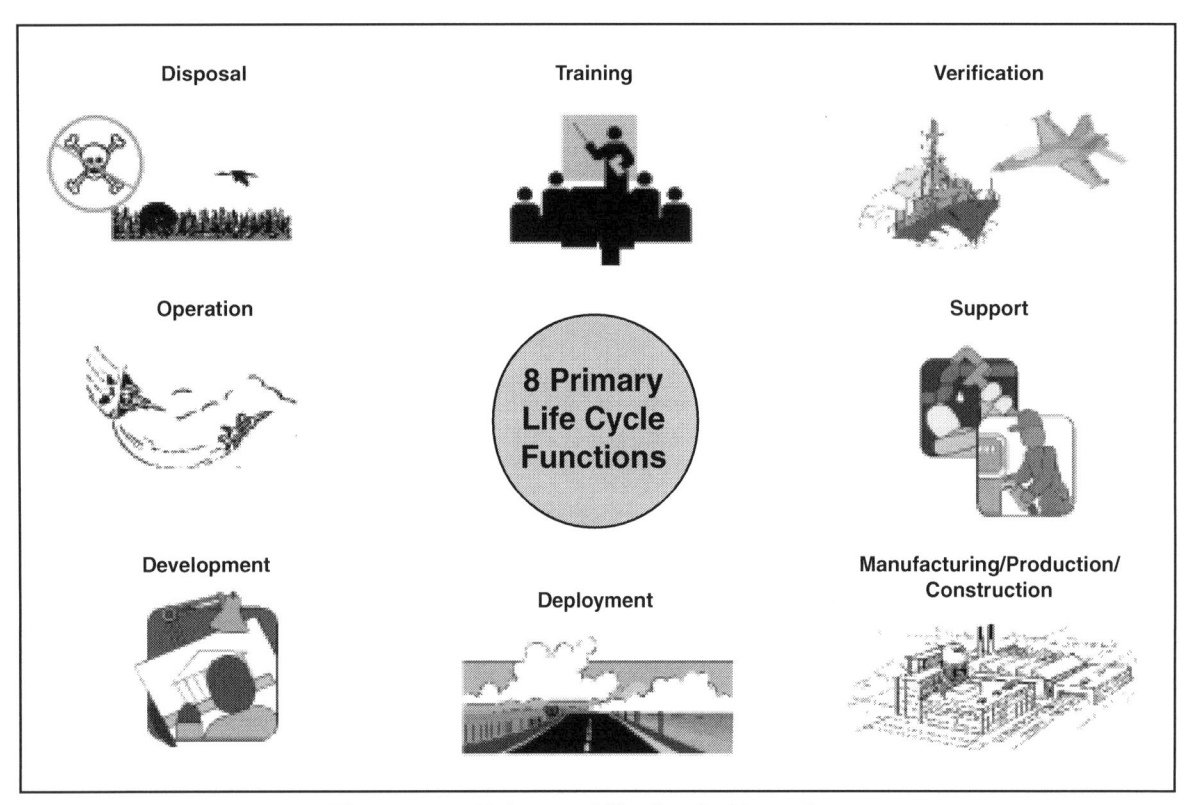

Figure 1-4. Primary Life Cycle Functions

Operation is the user function and includes activities necessary to satisfy defined operational objectives and tasks in peacetime and wartime environments.

Support includes the activities necessary to provide operations support, maintenance, logistics, and material management.

Disposal includes the activities necessary to ensure that the disposal of decommissioned, destroyed, or irreparable system components meets all applicable regulations and directives.

Training includes the activities necessary to achieve and maintain the knowledge and skill levels necessary to efficiently and effectively perform operations and support functions.

Verification includes the activities necessary to evaluate progress and effectiveness of evolving system products and processes, and to measure specification compliance.

Systems Engineering Considerations

Systems engineering is a standardized, disciplined management process for development of system solutions that provides a constant approach to system development in an environment of change and uncertainty. It also provides for simultaneous product and process development, as well as a common basis for communication.

Systems engineering ensures that the correct technical tasks get done during development through planning, tracking, and coordinating. Responsibilities of systems engineers include:

- Development of a total system design solution that balances cost, schedule, performance, and risk,

- Development and tracking of technical information needed for decision making,

- Verification that technical solutions satisfy customer requirements,

- Development of a system that can be produced economically and supported throughout the life cycle,

- Development and monitoring of internal and external interface compatibility of the system and subsystems using an open systems approach,

- Establishment of baselines and configuration control, and

- Proper focus and structure for system and major sub-system level design IPTs.

1.5 GUIDANCE

DoD 5000.2-R establishes two fundamental requirements for program management:

- It requires that an Integrated Product and Process approach be taken to design wherever practicable, and

- It requires that a disciplined systems engineering process be used to translate operational needs and/or requirements into a system solution.

Tailoring the Process

System engineering is applied during all acquisition and support phases for large- and small-scale systems, new developments or product improvements, and single and multiple procurements. The process must be tailored for different needs and/or requirements. Tailoring considerations include system size and complexity, level of system definition detail, scenarios and missions, constraints and requirements, technology base, major risk factors, and organizational best practices and strengths.

For example, systems engineering of software should follow the basic systems engineering approach as presented in this book. However, it must be tailored to accommodate the software development environment, and the unique progress tracking and verification problems software development entails. In a like manner, all technology domains are expected to bring their own unique needs to the process.

This book provides a conceptual-level description of systems engineering management. The specific techniques, nomenclature, and recommended methods are not meant to be prescriptive. Technical managers must tailor their systems engineering planning to meet their particular requirements and constraints, environment, technical domain, and schedule/budget situation.

However, the basic time-proven concepts inherent in the systems engineering approach must be retained to provide continuity and control. For complex system designs, a full and documented understanding of what the system must do should precede development of component performance descriptions, which should precede component detail descriptions. Though some parts of the system may be dictated as a constraint or interface, in general, solving the design problem should start with analyzing the requirements and determining what the system has to do before physical alternatives are chosen. Configurations must be controlled and risk must be managed.

Tailoring of this process has to be done carefully to avoid the introduction of substantial unseen risk and uncertainty. Without the control, coordination, and traceability of systems engineering, an environment of uncertainty results which will lead to surprises. Experience has shown that these surprises almost invariably lead to significant impacts to cost and schedule. Tailored processes that reflect the general conceptual approach of this book have been developed and adopted by professional societies, academia, industry associations, government agencies, and major companies.

1.6 SUMMARY POINTS

- Systems engineering management is a multi-functional process that integrates life cycle functions, the systems engineering problem-solving process, and progressive baselining.

- The systems engineering process is a problem-solving process that drives the balanced development of system products and processes.

- Integrated Product Teams should apply the systems engineering process to develop a life cycle balanced-design solution.

- The systems engineering process is applied to each level of development, one level at a time.

- Fundamental systems engineering activities are Requirements Analysis, Functional Analysis/Allocation, and Design Synthesis, all of which are balanced by System Analysis and Control.

- Baseline phasing provides for an increasing level of descriptive detail of the products and processes with each application of the systems engineering process.

- Baselining in a nut shell is a concept description that leads to a system definition which, in turn, leads to component definitions, and then to component designs, which finally lead to a product.

- The output of each application of the systems engineering process is a major input to the next process application.

CHAPTER 2

SYSTEMS ENGINEERING MANAGEMENT IN DOD ACQUISITION

2.1 INTRODUCTION

The DoD acquisition process has its foundation in federal policy and public law. The development, acquisition, and operation of military systems is governed by a multitude of public laws, formal DoD directives, instructions and manuals, numerous Service and Component regulations, and many inter-service and international agreements.

Managing the development and fielding of military systems requires three basic activities: technical management, business management, and contract management. As described in this book, systems engineering management is the technical management component of DoD acquisition management.

The acquisition process runs parallel to the requirements generation process and the budgeting process (Planning, Programming, and Budgeting System.) User requirements tend to be event driven by threat. The budgeting process is date driven by constraints of the Congressional calendar. Systems Engineering Management bridges these processes and must resolve the dichotomy of event driven needs, event driven technology development, and a calendar driven budget.

Direction and Guidance

The Office of Management and Budget (OMB) provides top-level guidance for planning, budgeting, and acquisition in OMB Circular A-11, Part 3, and the Supplemental Capital Programming Guide: Planning, Budgeting, and Acquisition of Capital Assets, July 1997. These documents establish the broad responsibilities and ground rules to be followed in funding and acquiring major assets. The departments of the executive branch of government are then expected to draft their own guidance consistent with the guidelines established. The principal guidance for defense system acquisitions is the DoD 5000 series of directives and regulations. These documents reflect the actions required of DoD acquisition managers to:

- Translate operational needs into stable, affordable programs,

- Acquire quality products, and

- Organize for efficiency and effectiveness.

2.2 RECENT CHANGES

The DoD 5000 series documents were revised in 2000 to make the process more flexible, enabling the delivery of advanced technology to warfighters more rapidly and at reduced total ownership cost. The new process encourages multiple entry points, depending on the maturity of the fundamental technologies involved, and the use of evolutionary methods to define and develop systems. This encourages a tailored approach to acquisition and engineering management, but it does not alter the basic logic of the underlying systems engineering process.

2.3 ACQUISITION LIFE CYCLE

The revised acquisition process for major defense systems is shown in Figure 2-1. The process is

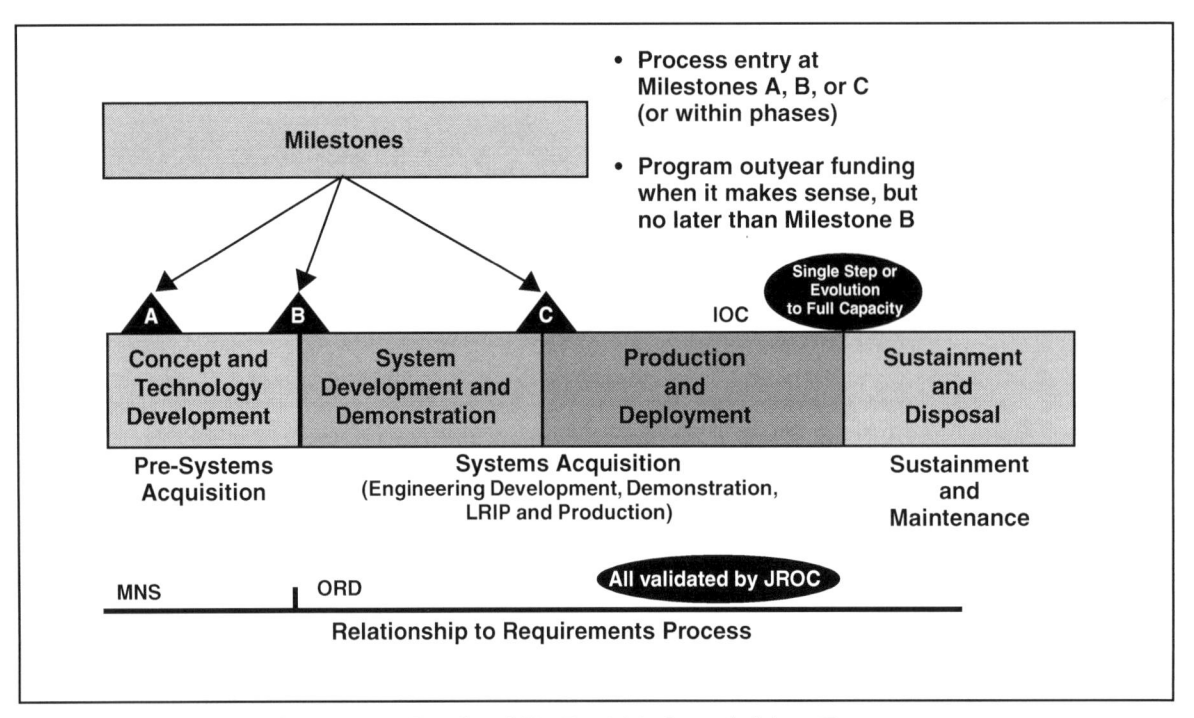

Figure 2-1. Revised DoD 5000 Acquisition Process

defined by a series of phases during which technology is defined and matured into viable concepts, which are subsequently developed and readied for production, after which the systems produced are supported in the field.

The process allows for a given system to enter the process at any of the development phases. For example, a system using unproven technology would enter at the beginning stages of the process and would proceed through a lengthy period of technology maturation, while a system based on mature and proven technologies might enter directly into engineering development or, conceivably, even production. The process itself (Figure 2-1) includes four phases of development. The first, **Concept and Technology Development**, is intended to explore alternative concepts based on assessments of operational needs, technology readiness, risk, and affordability. Entry into this phase *does not* imply that DoD has committed to a new acquisition program; rather, it is the initiation of a process to determine whether or not a need (typically described in a Mission Need Statement (MNS)) can be met at reasonable levels of technical risk and at affordable costs. The decision to enter into the Concept and Technology Development phase is made formally at the Milestone A forum.

The **Concept and Technology Development** phase begins with concept exploration. During this stage, concept studies are undertaken to define alternative concepts and to provide information about capability and risk that would permit an objective comparison of competing concepts. A decision review is held after completion of the concept exploration activities. The purpose of this review is to determine whether further technology development is required, or whether the system is ready to enter into system acquisition. If the key technologies involved are reasonably mature and have already been demonstrated, the Milestone Decision Authority (MDA) may agree to allow the system to proceed into system acquisition; if not, the system may be directed into a component advanced development stage. (See Supplement A to this chapter for a definition of Technology Readiness levels.) During this stage, system architecture definition will continue and key technologies will be demonstrated in order to ensure that technical and cost risks are understood and are at acceptable levels prior to entering acquisition. In any event, the

Concept and Technology Development phase ends with a defined system architecture supported by technologies that are at acceptable levels of maturity to justify entry into system acquisition.

Formal system acquisition begins with a Milestone B decision. The decision is based on an integrated assessment of technology maturity, user requirements, and funding. A successful Milestone B is followed by the **System Development and Demonstration** phase. This phase could be entered directly as a result of a technological opportunity and urgent user need, as well as having come through concept and technology development. The System Development and Demonstration phase consists of two stages of development, system integration and system demonstration. Depending upon the maturity level of the system, it could enter at either stage, or the stages could be combined. This is the phase during which the technologies, components and subsystems defined earlier are first integrated at the system level, and then demonstrated and tested. If the system has never been integrated into a complete system, it will enter this phase at the system integration stage. When subsystems have been integrated, prototypes demonstrated, and risks are considered acceptable, the program will normally enter the system demonstration stage following an interim review by the MDA to ensure readiness. The system demonstration stage is intended to demonstrate that the system has operational utility consistent with the operational requirements. Engineering demonstration models are developed and system level development testing and operational assessments are performed to ensure that the system performs as required. These demonstrations are to be conducted in environments that represent the eventual operational environments intended. Once a system has been demonstrated in an operationally relevant environment, it may enter the Production and Deployment phase.

The **Production and Deployment** phase consists of two stages: production readiness and low rate initial production (LRIP), and rate production and deployment. The decision forum for entry into this phase is the Milestone C event. Again, the fundamental issue as to where a program enters the process is a function of technology maturity, so the possibility exists that a system could enter directly into this phase if it were sufficiently mature, for example, a commercial product to be produced for defense applications. However the entry is made—directly or through the maturation process described, the production readiness and LRIP stage is where initial operational test, live fire test, and low rate initial production are conducted. Upon completion of the LRIP stage and following a favorable Beyond LRIP test report, the system enters the rate production and deployment stage during which the item is produced and deployed to the user. As the system is produced and deployed, the final phase, Sustainment and Disposal, begins.

The last, and longest, phase is the **Sustainment and Disposal** phase of the program. During this phase all necessary activities are accomplished to maintain and sustain the system in the field in the most cost-effective manner possible. The scope of activities is broad and includes everything from maintenance and supply to safety, health, and environmental management. This period may also include transition from contractor to organic support, if appropriate. During this phase, modifications and product improvements are usually implemented to update and maintain the required levels of operational capability as technologies and threat systems evolve. At the end of the system service life it is disposed of in accordance with applicable classified and environmental laws, regulations, and directives. Disposal activities also include recycling, material recovery, salvage of reutilization, and disposal of by-products from development and production.

The key to this model of the acquisition process is that programs have the flexibility to enter at any of the first three phases described. The decision as to where the program should enter the process is primarily a function of user needs and technology maturity. The MDA makes the decision for the program in question. Program managers are encouraged to work with their users to develop evolutionary acquisition strategies that will permit deliveries of usable capabilities in as short a time-frame as possible, with improvements and enhancements added as needed through continuing definition of requirements and development activities to support the evolving needs.

2.4 SYSTEMS ENGINEERING IN ACQUISITION

As required by DoD 5000.2-R, the systems engineering process shall:

1. Transform operational needs and requirements into an integrated system design solution through concurrent consideration of all life-cycle needs (i.e., development, manufacturing, test and evaluation, verification, deployment, operations, support, training and disposal).

2. Ensure the compatibility, interoperability and integration of all functional and physical interfaces and ensure that system definition and design reflect the requirements for all system elements: hardware, software, facilities, people, and data; and

3. Characterize and manage technical risks.

4. Apply scientific and engineering principles to identify security vulnerabilities and to minimize or contain associated information assurance and force protection risks.

These objectives are accomplished with use of the management concepts and techniques described in the chapters which follow in this book. The application of systems engineering management coincides with acquisition phasing. In order to support milestone decisions, major technical reviews are conducted to evaluate system design maturity.

Concept and Technology Development

The Concept and Technology Development phase consists of two pre-acquisition stages of development. The first, **Concept Exploration**, is represented in Figure 2-2. The exploration of concepts is usually accomplished through multiple short-term studies. Development of these studies is

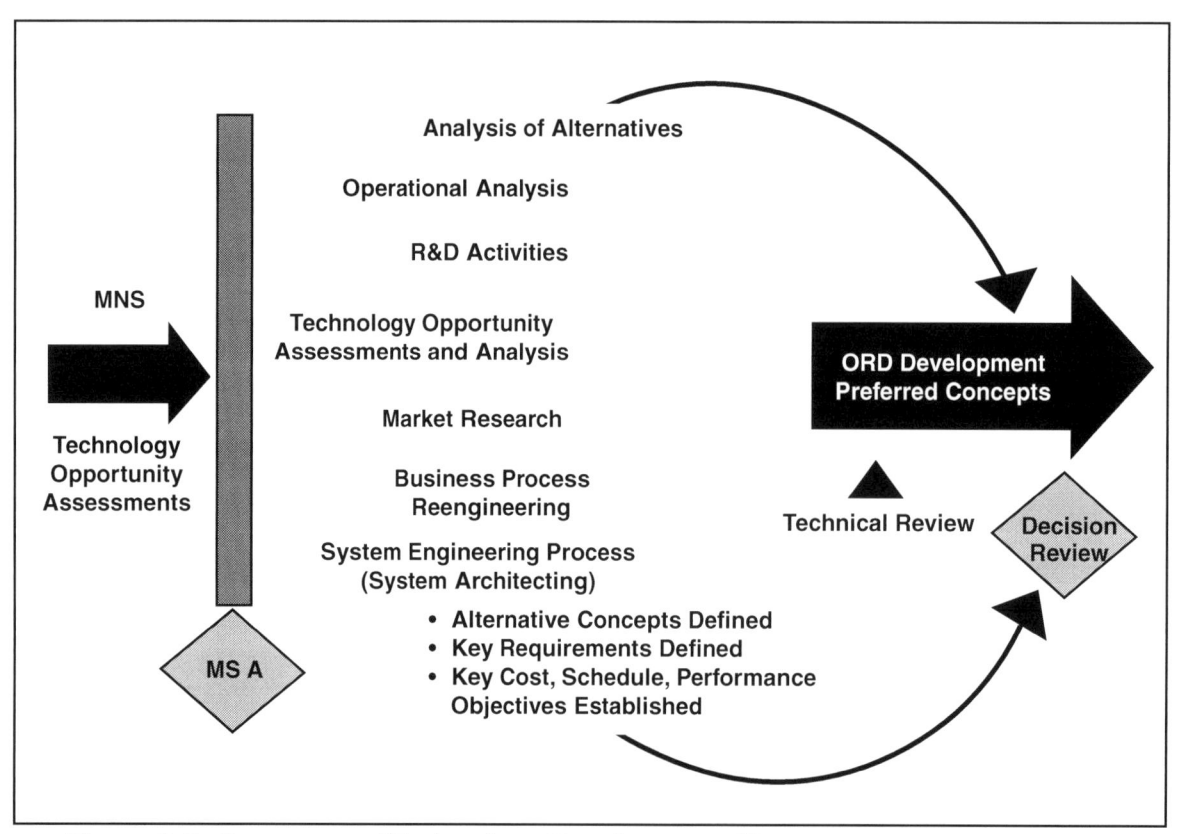

Figure 2-2. Concept and Technology Development (Concept Exploration Stage)

expected to employ various techniques including the systems engineering process that translates inputs into viable concept architectures whose functionality can be traced to the requirements. In addition, market surveys, Business Process Reengineering activities, operational analysis, and trade studies support the process.

The primary inputs to these activities include requirements, in form of the MNS, assessments of technology opportunities and status, and the outputs from any efforts undertaken to explore potential solutions. When the contractor studies are complete, a specific concept to be pursued is selected based on a integrated assessment of technical performance; technical, schedule and cost risk; as well as other relevant factors. A decision review is then held to evaluate the concept recommended and the state of technology upon which the concept depends. The MDA then makes a decision as to whether the concept development work needs to be extended or redirected, or whether the technology maturity is such that the program can proceed directly to either Mile-stone B (System Development and Demonstration) or Milestone C (Production and Deployment).

If the details of the concept require definition, i.e., the system has yet to be designed and demonstrated previously, or the system appears to be based on technologies that hold significant risk, then it is likely that the system will proceed to the second stage of the Concept and Technology Development phase. This stage, **Component Advanced Development**, is represented in Figure 2-3. This is also a pre-acquisition stage of development and is usually characterized by extensive involvement of the DoD Science and Technology (S&T) community. The fundamental objectives of this stage of development are to define a system-level architecture and to accomplish risk-reduction activities as required to establish confidence that the building blocks of the system are sufficiently well-defined, tested and demonstrated to provide confidence that, when integrated into higher level assemblies and subsystems, they will perform reliably.

Development of a system-level architecture entails continuing refinement of system level requirements based on comparative analyses of the system concepts under consideration. It also requires that consideration be given to the role that the system

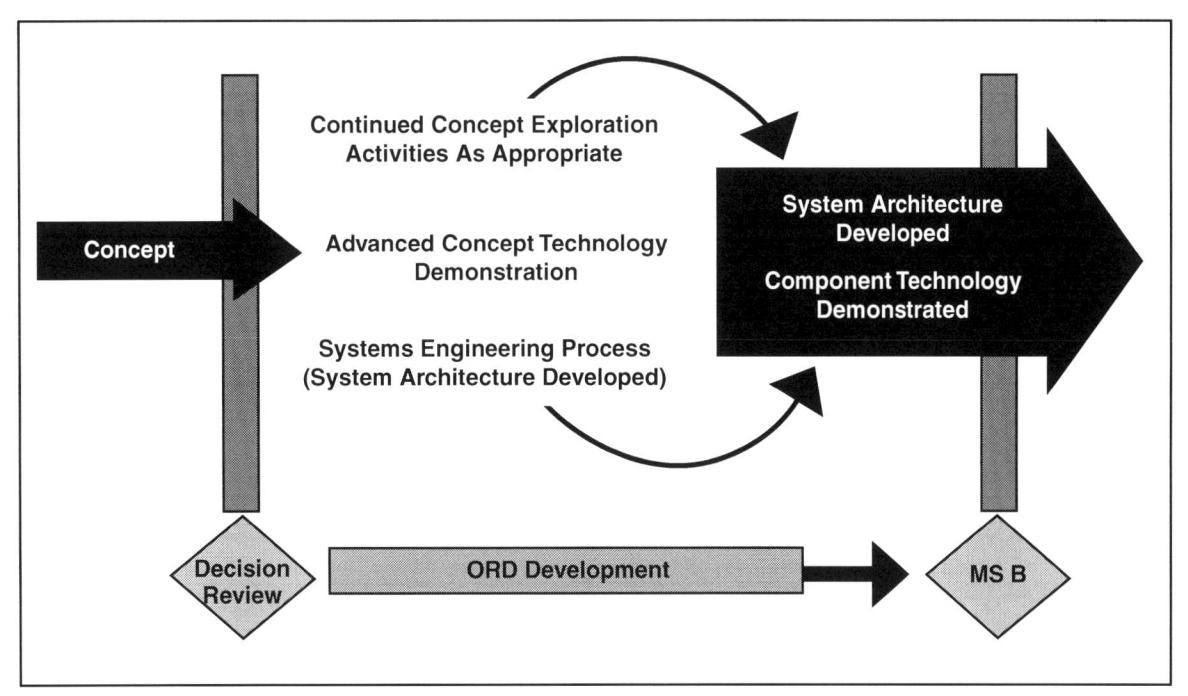

Figure 2-3. Concept and Technology Development (Component Advanced Development Stage)

will play in the system of systems of which it will be a part. System level interfaces must be established. Communications and interoperability requirements must be established, data flows defined, and operational concepts refined. Top level planning should also address the strategies that will be employed to maintain the supportability and affordability of the system over its life cycle including the use of common interface standards and open systems architectures. Important design requirements such as interoperability, open systems, and the use of commercial components should also be addressed during this stage of the program.

Risk reduction activities such as modeling and simulation, component testing, bench testing, and man-in-the-loop testing are emphasized as decisions are made regarding the various technologies that must be integrated to form the system. The primary focus at this stage is to ensure that the key technologies that represent the system components (assemblies and sub-systems) are well understood and are mature enough to justify their use in a system design and development effort. The next stage of the life cycle involves engineering development, so research and development (R&D) activities conducted within the science and technology appropriations should be completed during this stage.

System Development and Demonstration

The decision forum for entry into the System Development and Demonstration (SD&D) phase is the Milestone B event. Entry into this phase represents *program initiation*, the formal beginning of a system acquisition effort. This is the government commitment to pursue the program. Entry requires mature technology, validated requirements, and funding. At this point, the program requirement must be defined by an Operational Requirements Document (ORD). This phase consists of two primary stages, system integration (Figure 2-4) and system demonstration (Figure 2-5).

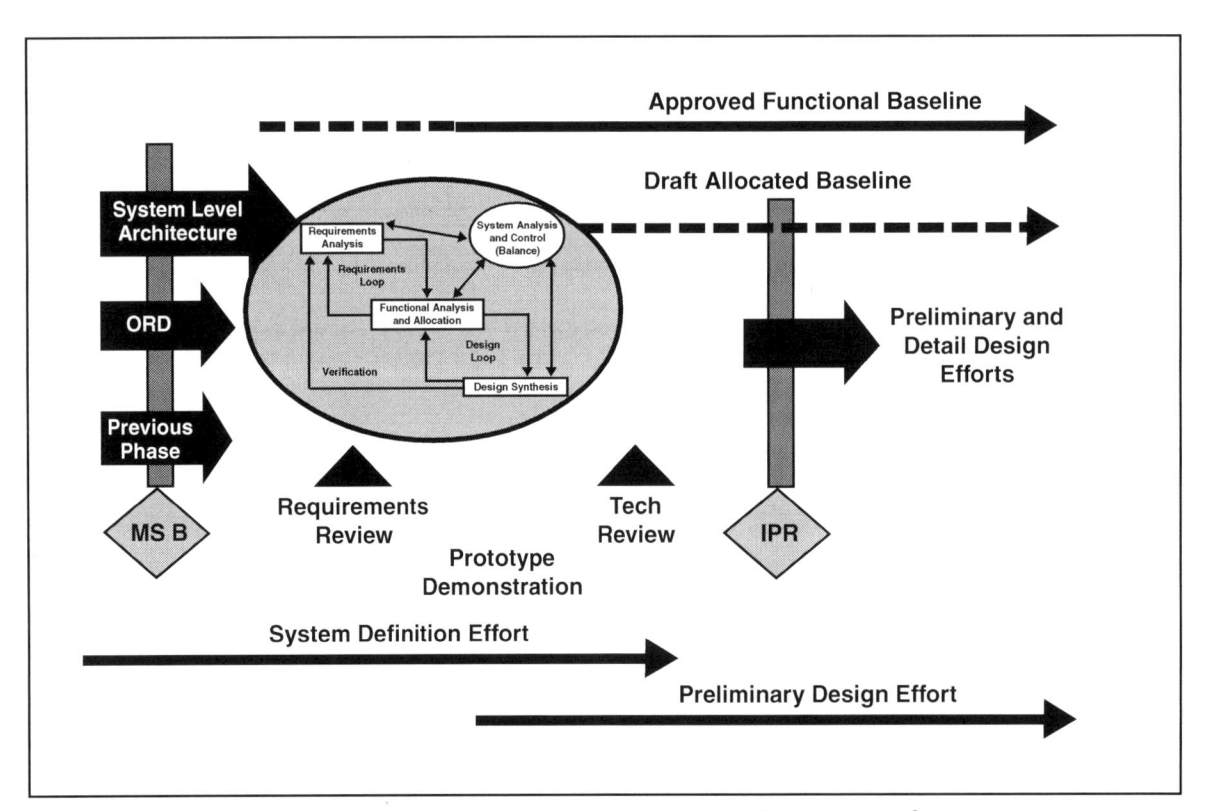

Figure 2-4. System Development and Demonstration (System Integration Stage)

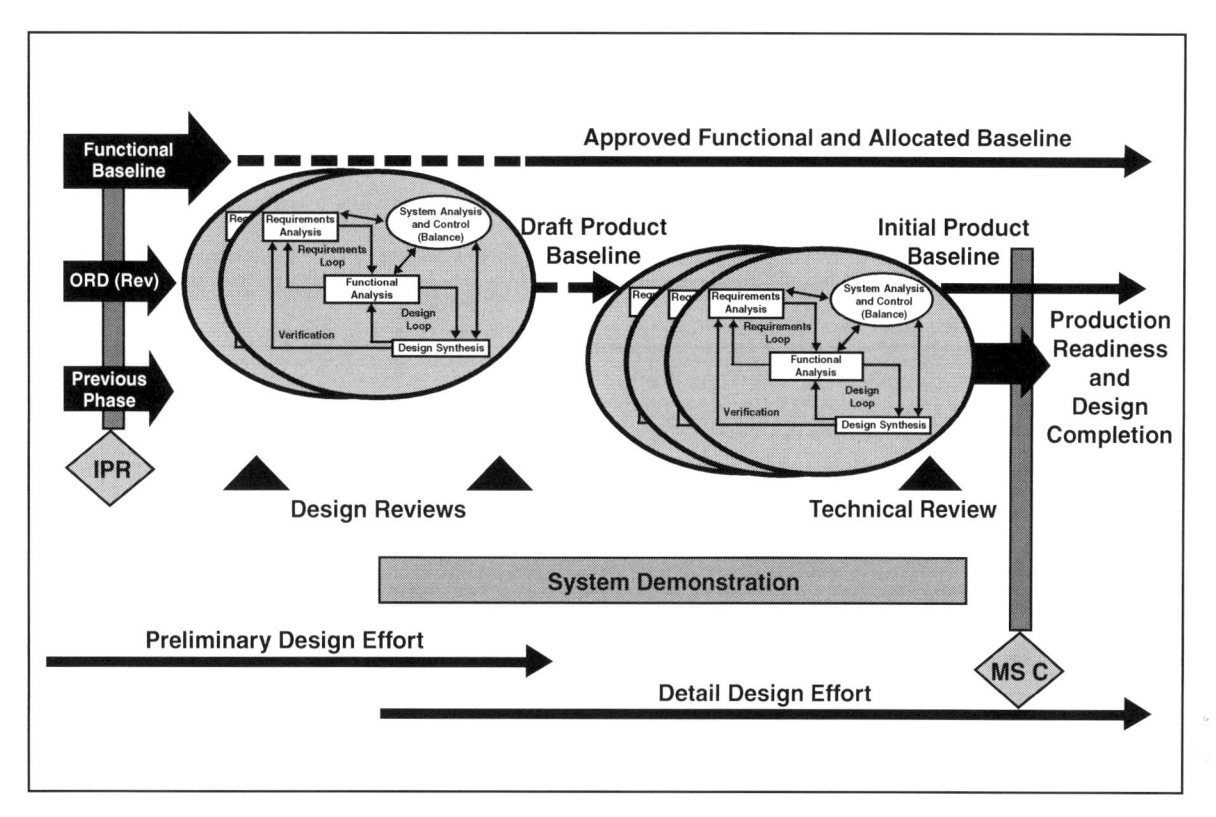

Figure 2-5. System Development and Demonstration (System Demonstration Stage)

There is no hard and fast guidance that stipulates precisely how the systems engineering process is to intersect with the DoD acquisition process. There are no specified technical events, e.g., DoD designated technical reviews, that are to be accomplished during identified stages of the SD&D phase. However, the results of a SD&D phase should support a production go-ahead decision at Milestone C. That being the case, the process described below reflects a configuration control approach that includes a system level design (functional baseline), final preliminary designs (allocated baselines), and detail designs (initial product baselines). Along with their associated documentation, they represent the systems engineering products developed during SD&D that are most likely needed to appropriately support Milestone C.

System Integration is that stage of SD&D that applies to systems that have yet to achieve system level design maturity as demonstrated by the integration of components at the system level in relevant environments. For an unprecedented system (one not previously defined and developed), this stage will continue the work begun in the component advanced development stage, but the flavor of the effort now becomes oriented to engineering development, rather than the research-oriented efforts that preceded this stage. A formal ORD, technology assessments, and a high-level system architecture have been established. (These will form major inputs to the systems engineering process.) The engineering focus becomes establishment and agreement on system level technical requirements stated such that designs based on those technical requirements will meet the intent of the operational requirements. The system level technical requirements are stabilized and documented in an approved system level requirements specification. In addition, the system-level requirements baseline (functional baseline) is established. This baseline is verified by development and demonstration of prototypes that show that key technologies can be integrated and that associated risks are sufficiently low to justify developing the system.

Program initiation signals the transition from an S&T focus to management by the program office. The R&D community, the users, and the program office may have all been involved in defining the concepts and technologies that will be key to the system development. It is appropriate at this point, therefore, to conduct a thorough requirements analysis and review to ensure that the user, the contractor, and the program office all hold a common view of the requirements and to preserve the lessons learned through the R&D efforts conducted in the earlier phase. The risk at this point can be high, because misunderstandings and errors regarding system-level requirements will flow down to subsequent designs and can eventually result in overruns and even program failure. The contractor will normally use the occasion of the system requirements review early in this stage to set the functional baseline that will govern the flow-down of requirements to lower level items as preliminary designs are elaborated.

The Interim Progress Review held between System Integration and System Demonstration has no established agenda. The agenda is defined by the MDA and can be flexible in its timing and content. Because of the flexibility built into the acquisition process, not all programs will conform to the model presented here. Programs may find themselves in various stages of preliminary design and detailed design as the program passes from one stage of the SD&D phase to the succeeding stage. With these caveats, **System Demonstration** (Figure 2-5) is the stage of the SD&D phase during which preliminary and detailed designs are elaborated, engineering demonstration models are fabricated, and the system is demonstrated in operationally relevant environments.

System level requirements are flowed down to the lower level items in the architecture and requirements are documented in the item performance specifications, which represent the preliminary design requirements for those items. The item performance specifications and supporting documentation, when finalized, together form the allocated baseline for the system. Design then proceeds toward the elaboration of a detailed design for the product or system. The product baseline is drafted as the design is elaborated. This physical description of the system may change as a result of testing that will follow, but it forms the basis for initial fabrication and demonstration of these items. If the system has been previously designed and fabricated, then, clearly, this process would be curtailed to take advantage of work already completed.

Following the elaboration of the detailed design, components and subsystems are fabricated, integrated, and tested in a bottom-up approach until system level engineering demonstration models are developed. These demonstration models are not, as a rule, production representative systems. Rather, they are system demonstration models, or integrated commercial items, that serve the purpose of enabling the developer to accomplish development testing on the integrated system. These models are often configured specifically to enable testing of critical elements of the system, for example, in the case of an aircraft development, there may be separate engineering demonstration models developed specifically to test the integrated avionics subsystems, while others demonstrate the flying qualities and flight controls subsystems.

For purposes of making decisions relative to progress through the acquisition process, these system-level demonstrations are not intended to be restricted to laboratory test and demonstrations. They are expected to include rigorous demonstrations that the integrated system is capable of performing operationally useful tasks under conditions that, while not necessarily equal to the rigor of formal operational testing, represent the eventual environment in which the system must perform. The result of these demonstrations provide the confidence required to convince the decision-maker (MDA) that the system is ready to enter the production phase of the life cycle. This implies that the system has demonstrated not only that technical performance is adequate, but also that the affordability, supportability, and producibility risks are sufficiently low to justify a production decision.

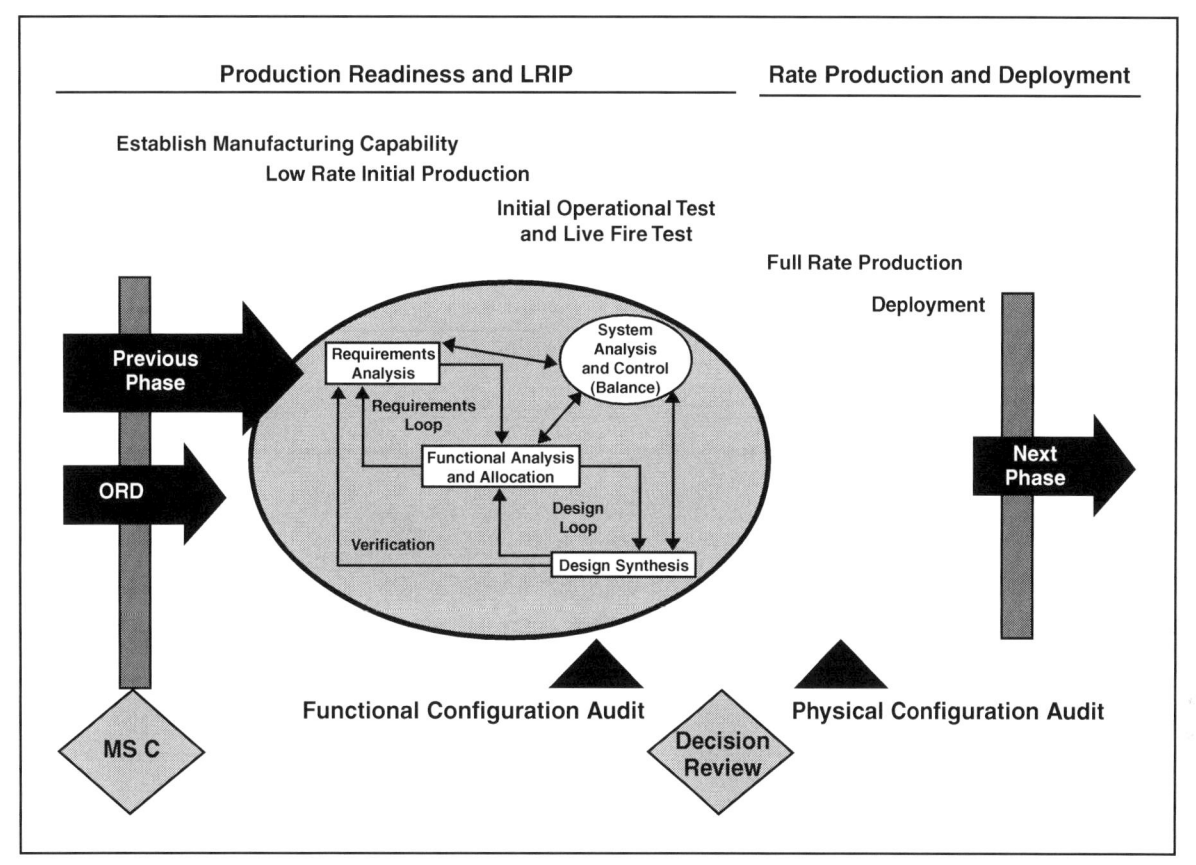

Figure 2-6. Production and Deployment

Production and Deployment

Milestone C is the decision forum for entry into the Production and Deployment phase of the program. Like other phases, this phase is also divided into stages of development. Production Readiness and LRIP is the first of these. At this point, system-level demonstrations have been accomplished and the product baseline is defined (although it will be refined as a result of the activities undertaken during this phase). The effort is now directed toward development of the manufacturing capability that will produce the product or system under development. When a manufacturing capability is established, a LRIP effort begins.

The development of a LRIP manufacturing capability has multiple purposes. The items produced are used to proof and refine the production line itself, items produced on this line are used for Initial Operational Test and Evaluation (IOT&E) and Live Fire Test and Evaluation (LFT&E), and this is also the means by which manufacturing rates are ramped upward to the rates intended when manufacturing is fully underway.

Following the completion of formal testing, the submission of required Beyond-LRIP and Live Fire Test reports, and a full-rate production decision by the MDA, the system enters the Rate Production and Deployment stage. After the decision to go to full-rate production, the systems engineering process is used to refine the design to incorporate findings of the independent operational testing, direction from the MDA, and feedback from deployment activities. Once configuration changes have been made and incorporated into production, and the configuration and production is considered stable, Follow-on Operational Test and Evaluation (FOT&E), if required, is typically performed on the stable production system. Test results are used to further refine the production configuration. Once this has been accomplished and production again becomes stable, detailed audits are held to confirm that the Product Baseline documentation correctly describes the system being produced. The Product Baseline is then put under formal configuration control.

As the system is produced, individual items are delivered to the field units that will actually employ and use them in their military missions. Careful coordination and planning is essential to make the deployment as smooth as possible. Integrated planning is absolutely critical to ensure that the training, equipment, and facilities that will be required to support the system, once deployed, are in place as the system is delivered. The systems engineering function during this activity is focused on the integration of the functional specialties to make certain that no critical omission has been made that will render the system less effective than it might otherwise be. Achieving the user's required initial operational capability (IOC) schedule demands careful attention to the details of the transition at this point. Furthermore, as the system is delivered and operational capability achieved, the system transitions to the Sustainment and Disposal phase of the system life cycle—the longest and most expensive of all phases.

Sustainment and Disposal

There is no separate milestone decision required for a program to enter this phase of the system life cycle. The requirement for the Sustainment phase is implicit in the decision to produce and deploy the system. This phase overlaps the Production phase. Systems Engineering activities in the Sustainment phase are focused on maintaining the system's performance capability relative to the threat the system faces. If the military threat changes or a technology opportunity emerges, then the system may require modification. These modifications must be approved at an appropriate level for the particular change being considered. The change then drives the initiation of new systems engineering processes, starting the cycle (or parts of it) all over again.

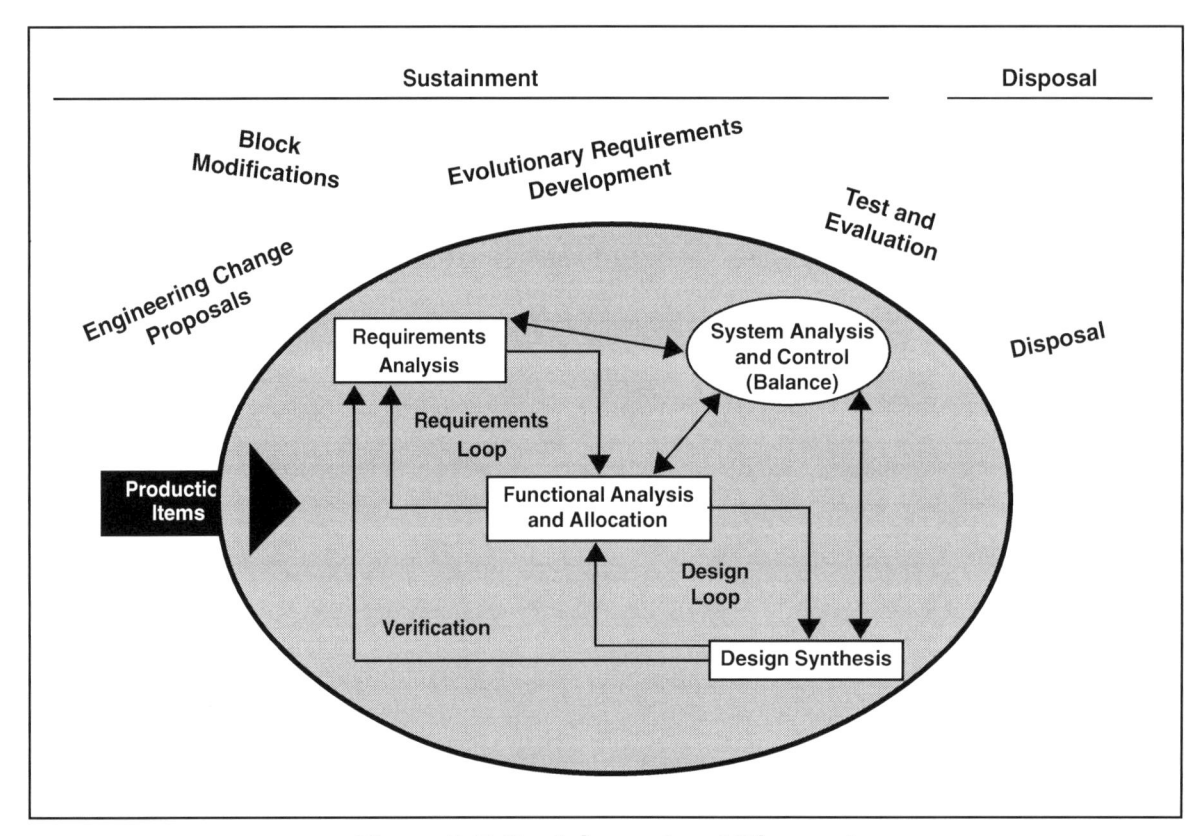

Figure 2-7. Sustainment and Disposal

Also, in an evolutionary development environment, there will be a continuing effort to develop and refine additional operational requirements based on the experience of the user with the portion of the system already delivered. As new requirements are generated, a new development cycle begins, with technology demonstrations, risk reduction, system demonstrations and testing—the same cycle just described—all tailored to the specific needs and demands of the technology to be added to the core system already delivered.

The final activity in the system life cycle is Disposal. System engineers plan for and conduct system disposal throughout the life cycle beginning with concept development. System components can require disposal because of decommissioning, their destruction, or irreparable damage. In addition, processes and material used for development, production, operation, or maintenance can raise disposal issues throughout the life cycle. Disposal must be done in accordance with applicable laws, regulations, and directives that are continually changing, usually to require more severe constraints. They mostly relate to security and environment issues that include recycling, material recovery, salvage, and disposal of by-products from development and production.

Every Development is Different

The process described above is intended to be very flexible in application. There is no "typical" system acquisition. The process is therefore defined to accommodate a wide range of possibilities, from systems that have been proven in commercial applications and are being purchased for military use, to systems that are designed and developed essentially from scratch. The path that the system development takes through the process will depend primarily on the level of maturity of the technology employed. As explained in the preceding discussion, if the system design will rely significantly on the use of proven or commercial items, then process can be adjusted to allow the system to skip phases, or move quickly from stage to stage within phases. If the type of system is well understood within the applicable technical domains, or it is an advanced version of a current well understood system, then the program definition and risk reduction efforts could be adjusted appropriately.

It is the role of the system engineer to advise the program manager of the recommended path that the development should take, outlining the reasons for that recommendation. The decision as to the appropriate path through the process is actually made by the MDA, normally based on the recommendation of the program manager. The process must be tailored to the specific development, both because it is good engineering and because it is DoD policy as part of the Acquisition Reform initiative. But tailoring must done with the intent of preserving the requirements traceability, baseline control, lifecycle focus, maturity tracking, and integration inherent in the systems engineering approach. The validity of tailoring the process should always be a risk management issue. Acquisition Reform issues will be addressed again in Part IV of this text.

2.5 SUMMARY POINTS

- The development, acquisition, and operation of military systems is governed by a multitude of public laws, formal DoD directives, instructions and manuals, numerous Service and Component regulations, and many inter-service and international agreements.

- The system acquisition life cycle process is a model used to guide the program manager through the process of maturing technology based systems and readying them for production and deployment to military users.

- The acquisition process model is intended to be flexible and to accommodate systems and technologies of varying maturities. Systems dependent on immature technologies will take longer to develop and produce, while those that employ mature technologies can proceed through the process relatively quickly.

- The system engineering effort is integrated into the systems acquisition process such that the activities associated with systems engineering (development of documentation, technical reviews, configuration management, etc.) support and strengthen the acquisition process. The challenge for the engineering manager is to ensure that engineering activities are conducted at appropriate points in the process to ensure that the system has, in fact, achieved the levels of maturity expected prior to progressing into succeeding phases.

SUPPLEMENT 2-A

TECHNOLOGY READINESS LEVELS

Technology Readiness Level	Description
1. Basic principles observed and reported.	Lowest level of technology readiness. Scientific research begins to be translated into technology's basic properties.
2. Technology concept and/or application formulated.	Invention begins. Once basic principles are observed, practical applications can be invented. The application is speculative and there is no proof or detailed analysis to support the assumption. Examples are still limited to paper studies.
3. Analytical and experimental critical function and/or characteristic proof of concept.	Active R&D is initiated. This includes analytical studies and laboratory studies to physically validate analytical predictions of separate elements of the technology. Examples include components that are not yet integrated or representative.
4. Component and/or breadboard validation in laboratory environment.	Basic technological components are integrated to establish that the pieces will work together. This is relatively "low fidelity" compared to the eventual system. Examples include integration of "ad hoc" hardware in a laboratory.
5. Component and/or breadboard validation in relevant environment.	Fidelity of breadboard technology increases significantly. The basic technological components are integrated with reasonably realistic supporting elements so that the technology can be tested in simulated environment. Examples include "high fidelity" laboratory integration of components.
6. System/subsystem model or prototype demonstration in a relevant environment.	Representative model or prototype system, which is well beyond the breadboard tested for level 5, is tested in a relevant environment. Represents a major step up in a technology's demonstrated readiness. Examples include testing a prototype in a high fidelity laboratory environment or in a simulated operational environment.
7. System prototype demonstration in an operational environment.	Prototype near or at planned operational system. Represents a major step up from level 6, requiring the demonstration of an actual system prototype in an operational environment. Examples include testing the prototype in a test bed aircraft.

(continued)

Technology Readiness Level	Description
8. Actual system completed and qualified through test and demonstration.	Technology has been proven to work in its final form and under expected conditions. In almost all cases, this level represents the end of true system development. Examples include developmental test and evaluation of the system in its intended weapon system to determine if it meets design specifications.
9. Actual system proven through successful mission operations.	Actual application of the technology in its final form and under mission conditions, such as those encountered in operational test and evaluation. Examples include using the system under operational mission conditions.

SUPPLEMENT 2-B

EVOLUTIONARY ACQUISITION CONSIDERATIONS

The evolutionary approach to defense acquisition is the simple recognition that systems evolve as a result of changing user needs, technological opportunities, and knowledge gained in operation. Evolutionary Acquisition is not new to military systems. No naval ship in a class is the same; aircraft and vehicles have block changes designed to improve the design; variants of systems perform different missions; satellites have evolutionary improvements between the first and last launched; and due to fast evolving technology, computer resources and software systems are in constant evolution.

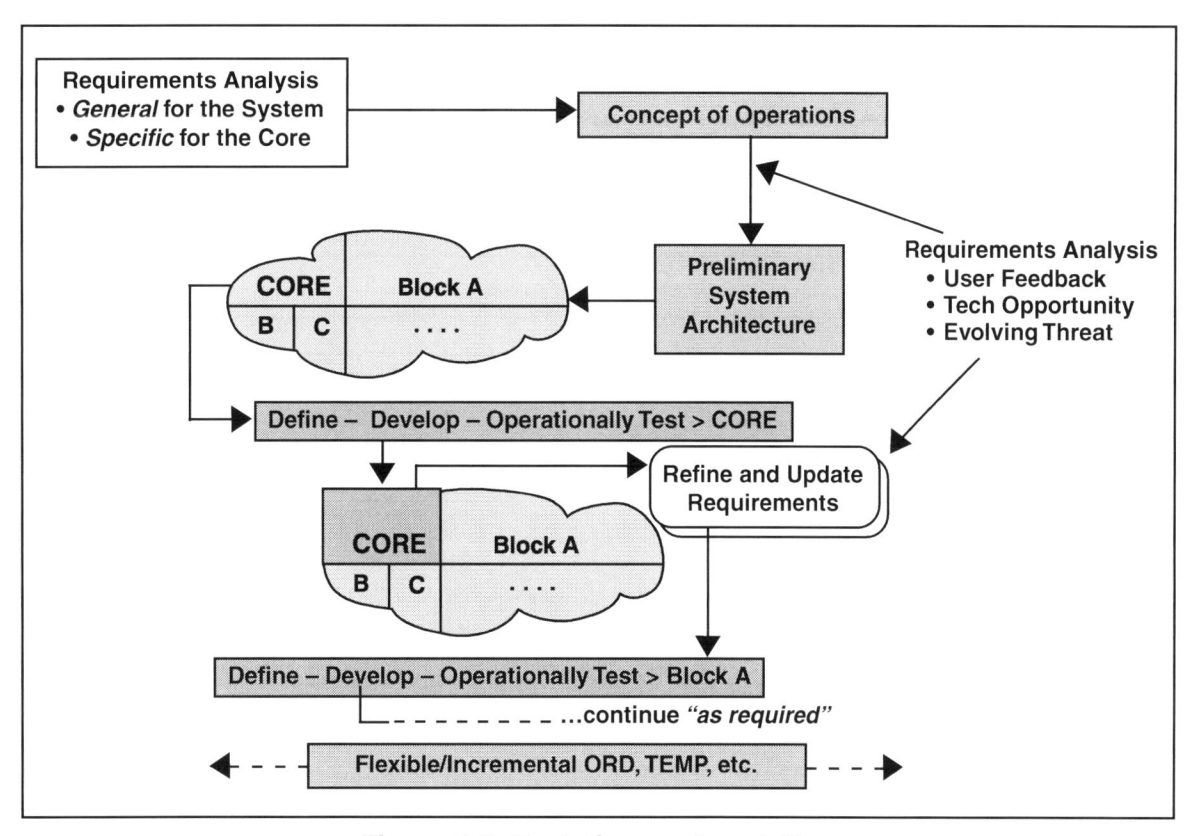

As shown by Figure 2-8, evolutionary acquisition starts with the development and delivery of a core capability. As knowledge is gained through system use and as technology changes, the system is evolved to a more useful or effective product. At the beginning of an evolutionary acquisition the ultimate user need is understood in general terms, but a core need that has immediate utility can be well-defined. Because future events will affect the eventual form of the product, the requirements can not be fully defined at the program initiation. However, the evolutionary development must be accomplished in a management system that demands

Figure 2-8. Evolutionary Acquisition

requirements validation, fully funded budgets, and rigorous review. In addition, the systems engineering function remains responsible for controlling requirements traceability and configuration control in the absence of complete definition of all requirements or final configurations. These constraints and concerns require the evolutionary approach be accomplished in a manner such the various concerns of users, developers, and managers are adequately addressed, while the risks associated with these issues are mitigated.

Acquisition Managment

Acquisition management requirements established in the DoD 5000 documents and associated component regulations or instructions establish a series of program-specific analyses, reports, and decision documents that support the milestone decision process. In addition, prior to decision points in the acquisition process, substantial coordination is required with an array of stakeholders. This process is resource consuming but necessary to establish the program's validity in the eyes of those responsible to approve the public resources committed to the program.

Evolutionary acquisition, by its nature, represents an "acquisition within an acquisition." On one level, the engineering manager is confronted with the management and control of the system as it progresses to its eventual final configuration, and, on another level, there is the management and control of the modifications, or blocks, that are successively integrated into the system as they are developed. The system has associated requirements, baselines, reviews—the normal elements of a system acquisition; however, each block also has specified requirements, configuration, and management activities. The challenge for technical management then becomes to ensure that good technical management principles are applied to the development of each block, while simultaneously ensuring that the definition and control of requirements and baselines at the system level include and accommodate the evolving architecture.

System Engineering Concerns

Evolutionary acquisition will require incremental and parallel development activities. These activities are developing evolutionary designs that represent a modification as well as an evolved system. The evolutionary upgrade is developed as a modification, but the new evolved system must be evaluated and verified as a system with new, evolved requirements. This implies that, though we can enter the acquisition process at any point, the basic baselining process required by systems engineering must somehow be satisfied for each block upgrade to assure requirements traceability and configuration control.

As shown by Figure 2-9, incremental delivery of capability can be the result of an evolutionary block upgrade or be an incremental release of capability within the approved program (or current evolutionary block) baseline. System engineering is concerned with both. There is no check list approach to structure these relationships, but the following is presented to provide some general guidance in a difficult and complex area of acquisition management planning and implementation.

Evolutionary upgrades may be based on known operational requirements where delivery of the capability is incremental due to immediate operational need, continuing refinement of the product baseline prior to full operational capability, and pre-planned parallel developments. If the modification is only at the allocated or product baseline, and the program's approved performance, cost, and schedule is not impacted, then the system would not necessarily require the management approvals and milestones normal to the acquisition process.

In all cases, the key to maintaining a proper systems engineering effort is to assure that architectures and configuration baselines used for evolutionary development can be upgraded with minimal impact to documented and demonstrated configurations. The risk associated with this issue can be significantly reduced through program planning that addresses optimization of the acquisition baseline and control of the evolving configuration.

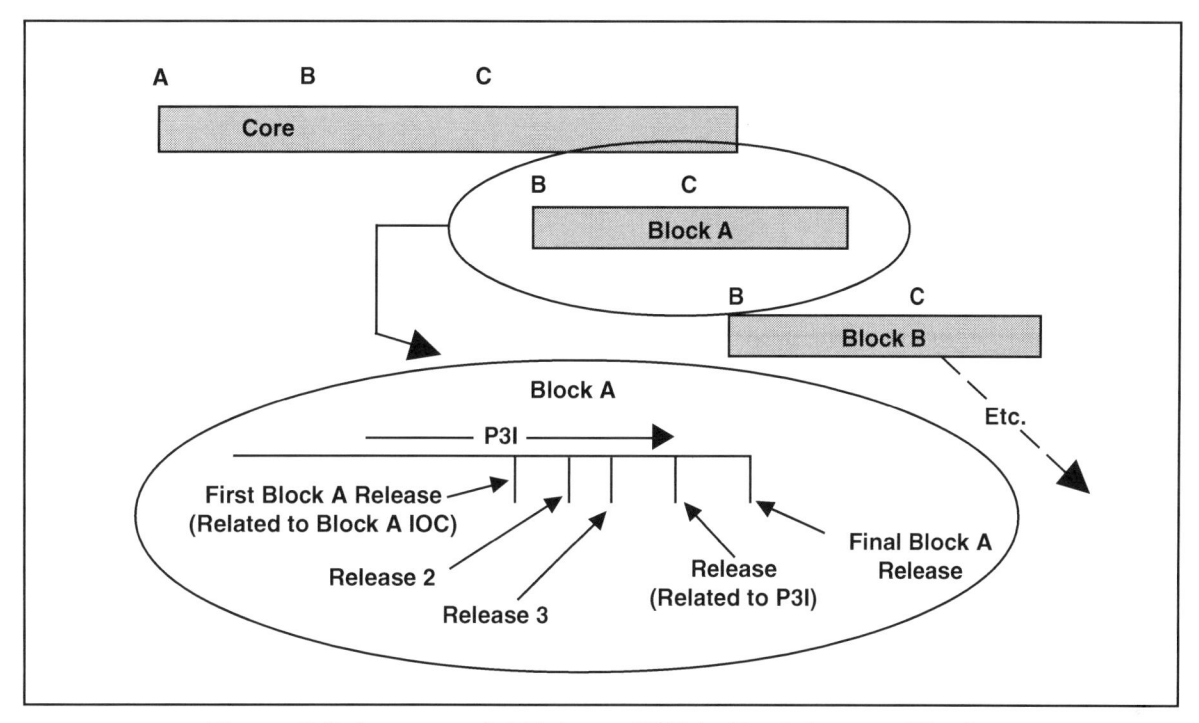

Figure 2-9. Incremental Release Within Evolutionary Blocks

Planning

Evolutionary acquisition program planning must clearly define how the core and evolutionary blocks will be structured, including:

1. A clear description of an operationally suitable core system including identification of subsystems and components most likely to evolve.

2. Establishment of a process for obtaining, evaluating and integrating operational feedback, technology advancements, and emerging commercial products.

3. Planning for evolutionary block upgrade evaluation, requirements validation, and program initiation.

4. Description of the management approach for evolutionary upgrades within a block and the constraints and controls associated with incremental delivery of capability.

5. Risk analysis of the developmental approach, both technical and managerial.

Systems engineering planning should emphasize:

1. The openness and modularity of the design of the core system architecture in order to facilitate modification and upgrades,

2. How baseline documentation is structured to improve flexibility for upgrade,

3. How evolutionary acquisition planning impacts baseline development and documentation control,

4. How technical reviews will be structured to best support the acquisition decision points, and

5. How risk management will monitor and control the management and technical complexity introduced by evolutionary development.

The basic system architecture should be designed to accommodate change. Techniques such as open architecting, functional partitioning, modular design, and open system design (all described later in this book) are key to planning for a flexible system that can be easily and affordably modified.

Notional Example of Evolutionary MAIS Acquisition Relationships					
Characterization	System Level	Acquisition Program Level	Acquisition Documentation Required	Baseline	CM Authority
Overall Need	Major Program or Business Area	Capstone or Sub-Portfolio	Capstone Acquisition Documentaion	Top Level Functional Baseline	PMO
Core and Evolutionary Blocks	Build or Block of Major Program	Acquisition Program	Full Program Documentation	Cumulative Functional and Allocated Baseline	PMO with Contractor Support
Incremental Delivery of Capability	Release or Version of Block	Internal to Acquisition Program	Separate Acquisition Documentation Not Required	Product Baseline	Contractor (Must Meet Allocated Basleine)
Associated Product Improvements	Application or Bridge	Parallel Product Improvement (Less than MAIS)	Component or Lower Decision Level Acquisition Processing	Functional, Allocated, and Product Baselines	PMO/Contractor

Table 2-1. Evolutionary Acquisition Relationships

Example

Table 2-1 illustrates some of the relationships discussed above as it might apply to a Major Automated Information System (MAIS) program. Due to the nature of complex software development, a MAIS acquisition inevitably will be an evolutionary acquisition. In the notional MAIS shown in the table, management control is primarily defined for capstone, program, subsystem or incremental delivery, and supporting program levels. The table provides relationships showing how key acquisition and system engineering activities correlate in the evolutionary environment. Probably the most important lesson of Table 2-1 is that these relationships are complex and if they are not planned for properly, they will present a significant risk to the program.

Summary

Acquisition oversight is directly related to the performance, cost, and schedule defined in the acquisition baseline. It establishes the approved scope of the developmental effort. Evolutionary development that exceeds the boundaries established by the acquisition baseline requires a new or revised acquisition review process with additional oversight requirements. The development and approval of the ORD and Acquisition Program Baseline are key activities that must structure an evolutionary process that provides user and oversight needs, budgetary control, requirements traceability, risk mitigation, and configuration management.

PART 2

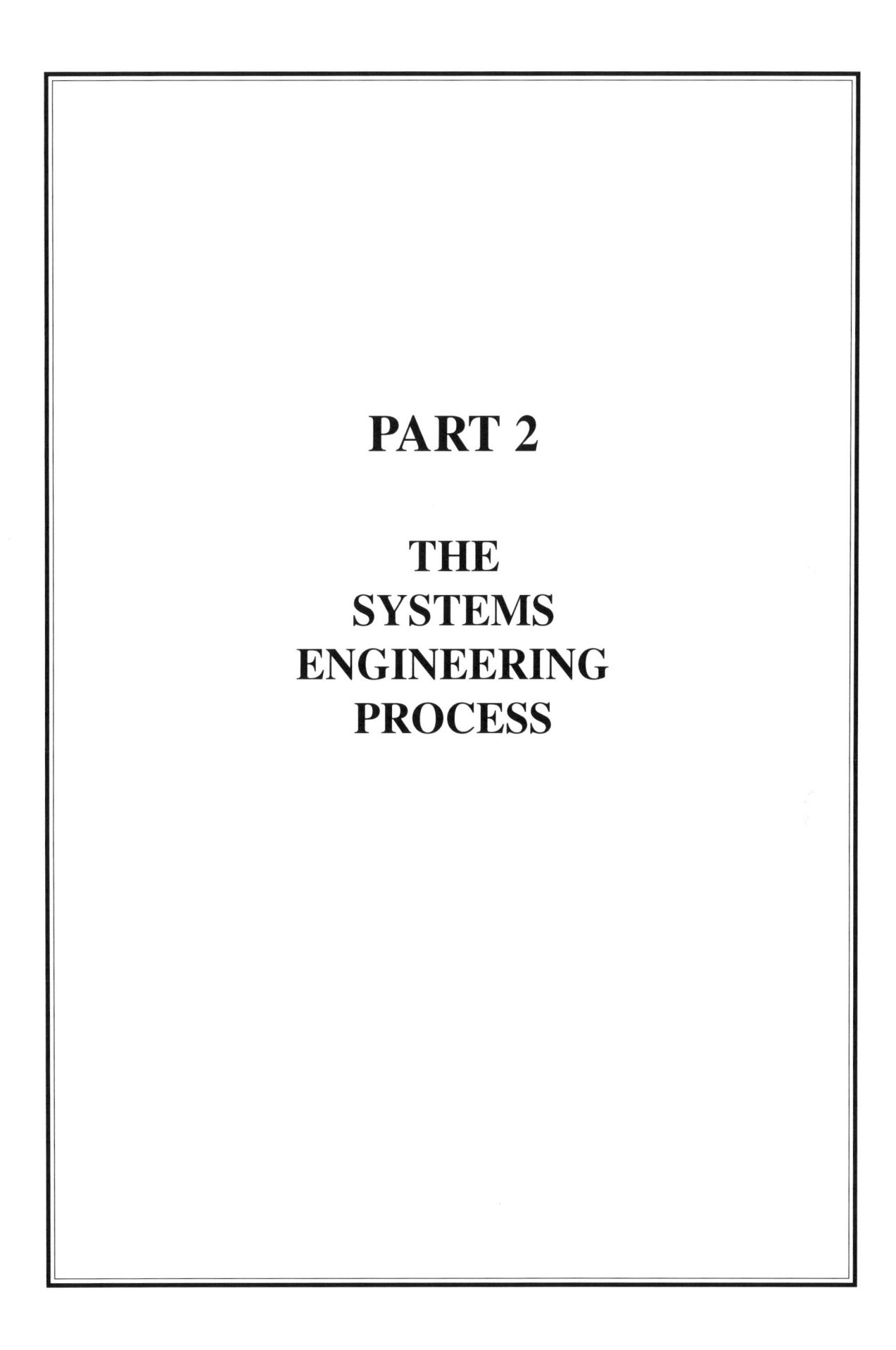

THE SYSTEMS ENGINEERING PROCESS

CHAPTER 3

SYSTEMS ENGINEERING PROCESS OVERVIEW

3.1 THE PROCESS

The Systems Engineering Process (SEP) is a comprehensive, iterative and recursive problem solving process, applied sequentially top-down by integrated teams. It transforms needs and requirements into a set of system product and process descriptions, generate information for decision makers, and provides input for the next level of development. The process is applied sequentially, one level at a time, adding additional detail and definition with each level of development. As shown by Figure 3-1, the process includes: inputs and outputs; requirements analysis; functional analysis and allocation; requirements loop; synthesis; design loop; verification; and system analysis and control.

Systems Engineering Process Inputs

Inputs consist primarily of the customer's needs, objectives, requirements and project constraints.

Process Input
- Customer Needs/Objectives/Requirements
  - Missions
  - Measures of Effectiveness
  - Environments
  - Constraints
- Technology Base
- Output Requirements from Prior Development Effort
- Program Decision Requirements
- Requirements Applied Through Specifications and Standards

Requirements Analysis
- Analyze Missions and Environments
- Identify Functional Requirements
- Define/Refine Performance and Design Constraint Requirements

Requirements Loop

Functional Analysis/Allocation
- Decompose to Lower-Level Functions
- Allocate Performance and Other Limiting Requirements to All Functional Levels
- Define/Refine Functional Interfaces (Internal/External)
- Define/Refine/Integrate Functional Architecture

Design Loop

Synthesis
- Transform Architectures (Functional to Physical)
- Define Alternative System Concepts, Configuration Items and System Elements
- Select Preferred Product and Process Solutions
- Define/Refine Physical Interfaces (Internal/External)

Verification

System Analysis and Control (Balance)
- Trade-Off Studies
- Effectiveness Analyses
- Risk Management
- Configuration Management
- Interface Management
- Data Management
- Perfomance Measurement
  - SEMS
  - TPM
  - Technical Reviews

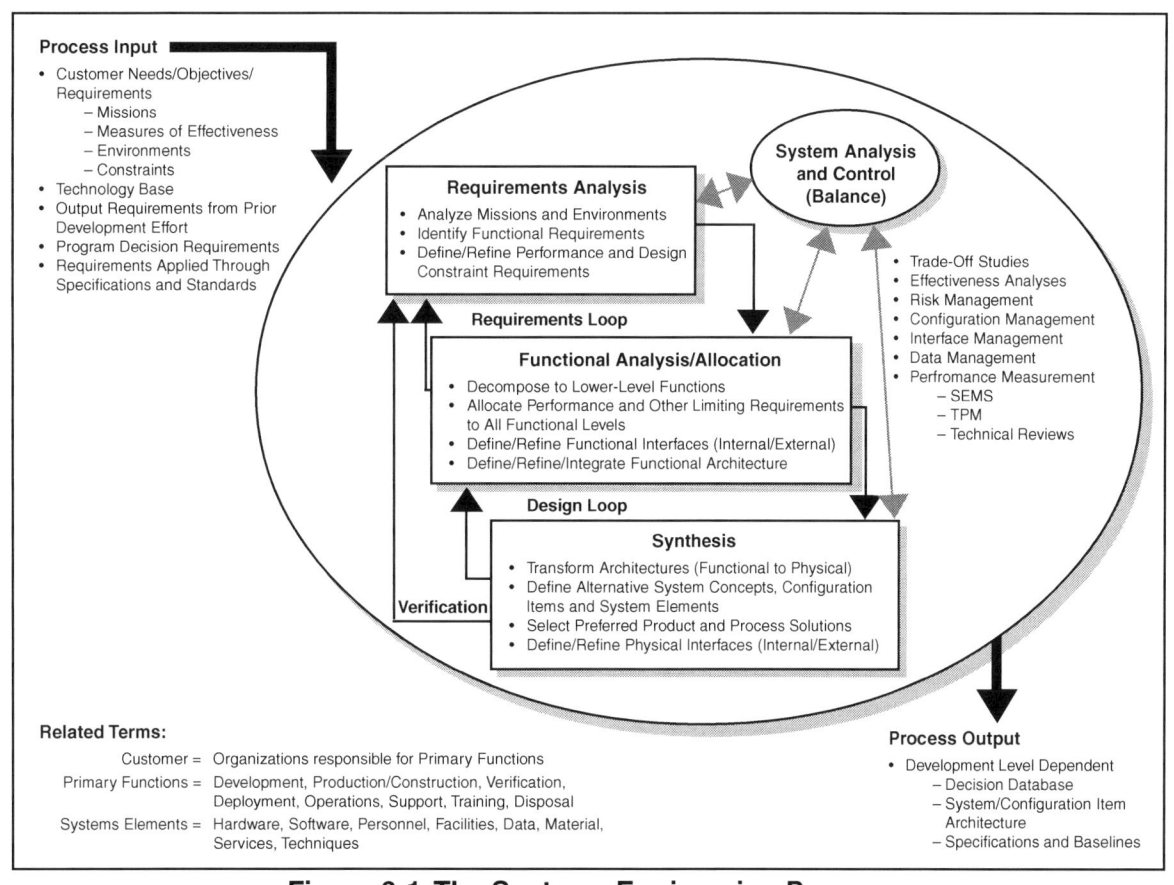

Process Output
- Development Level Dependent
  - Decision Database
  - System/Configuration Item Architecture
  - Specifications and Baselines

Related Terms:
Customer = Organizations responsible for Primary Functions
Primary Functions = Development, Production/Construction, Verification, Deployment, Operations, Support, Training, Disposal
Systems Elements = Hardware, Software, Personnel, Facilities, Data, Material, Services, Techniques

Figure 3-1. The Systems Engineering Process

Inputs can include, but are not restricted to, missions, measures of effectiveness, environments, available technology base, output requirements from prior application of the systems engineering process, program decision requirements, and requirements based on "corporate knowledge."

Requirements Analysis

The first step of the Systems Engineering Process is to analyze the process inputs. Requirements analysis is used to develop functional and performance requirements; that is, customer requirements are translated into a set of requirements that define what the system must do and how well it must perform. The systems engineer must ensure that the requirements are understandable, unambiguous, comprehensive, complete, and concise.

Requirements analysis must clarify and define functional requirements and design constraints. Functional requirements define quantity (how many), quality (how good), coverage (how far), time lines (when and how long), and availability (how often). Design constraints define those factors that limit design flexibility, such as: environmental conditions or limits; defense against internal or external threats; and contract, customer or regulatory standards.

Functional Analysis/Allocation

Functions are analyzed by decomposing higher-level functions identified through requirements analysis into lower-level functions. The performance requirements associated with the higher level are allocated to lower functions. The result is a description of the product or item in terms of what it does logically and in terms of the performance required. This description is often called the functional architecture of the product or item. Functional analysis and allocation allows for a better understanding of what the system has to do, in what ways it can do it, and to some extent, the priorities and conflicts associated with lower-level functions. It provides information essential to optimizing physical solutions. Key tools in functional analysis and allocation are Functional Flow Block Diagrams, Time Line Analysis, and the Requirements Allocation Sheet.

Requirements Loop

Performance of the functional analysis and allocation results in a better understanding of the requirements and should prompt reconsideration of the requirements analysis. Each function identified should be traceable back to a requirement. This iterative process of revisiting requirements analysis as a result of functional analysis and allocation is referred to as the requirements loop.

Design Synthesis

Design synthesis is the process of defining the product or item in terms of the physical and software elements which together make up and define the item. The result is often referred to as the physical architecture. Each part must meet at least one functional requirement, and any part may support many functions. The physical architecture is the basic structure for generating the specifications and baselines.

Design Loop

Similar to the requirements loop described above, the design loop is the process of revisiting the functional architecture to verify that the physical design synthesized can perform the required functions at required levels of performance. The design loop permits reconsideration of how the system will perform its mission, and this helps optimize the synthesized design.

Verification

For each application of the system engineering process, the solution will be compared to the requirements. This part of the process is called the verification loop, or more commonly, Verification. Each requirement at each level of development must be verifiable. Baseline documentation developed during the systems engineering process must establish the method of verification for each requirement.

Appropriate methods of verification include examination, demonstration, analysis (including modeling and simulation), and testing. Formal test and evaluation (both developmental and operational) are important contributors to the verification of systems.

Systems Analysis and Control

Systems Analysis and Control include technical management activities required to measure progress, evaluate and select alternatives, and document data and decisions. These activities apply to all steps of the sysems engineering process.

System analysis activities include trade-off studies, effectiveness analyses, and design analyses. They evaluate alternative approaches to satisfy technical requirements and program objectives, and provide a rigorous quantitative basis for selecting performance, functional, and design requirements. Tools used to provide input to analysis activities include modeling, simulation, experimentation, and test.

Control activities include risk management, configuration management, data management, and performance-based progress measurement including event-based scheduling, Technical Performance Measurement (TPM), and technical reviews.

The purpose of Systems Analysis and Control is to ensure that:

- Solution alternative decisions are made only after evaluating the impact on system effectiveness, life cycle resources, risk, and customer requirements,

- Technical decisions and specification requirements are based on systems engineering outputs,

- Traceability from systems engineering process inputs to outputs is maintained,

- Schedules for development and delivery are mutually supportive,

- Required technical disciplines are integrated into the systems engineering effort,

- Impacts of customer requirements on resulting functional and performance requirements are examined for validity, consistency, desirability, and attainability, and,

- Product and process design requirements are directly traceable to the functional and performance requirements they were designed to fulfill, and vice versa.

Systems Engineering Process Output

Process output is dependent on the level of development. It will include the decision database, the system or configuration item architecture, and the baselines, including specifications, appropriate to the phase of development. In general, it is any data that describes or controls the product configuration or the processes necessary to develop that product.

3.2 SUMMARY POINTS

- The system engineering process is the engine that drives the balanced development of system products and processes applied to each level of development, one level at a time.

- The process provides an increasing level of descriptive detail of products and processes with each system engineering process application. The output of each application is the input to the next process application.

CHAPTER 4

REQUIREMENTS ANALYSIS

4.1 SYSTEMS ENGINEERING PROCESS INPUTS

The inputs to the process include the customer's requirements and the project constraints. Requirements relate directly to the performance characteristics of the system being designed. They are the stated life-cycle customer needs and objectives for the system, and they relate to how well the system will work in its intended environment.

Constraints are conditions that exist because of limitations imposed by external interfaces, project support, technology, or life cycle support systems. Constraints bound the development teams' design opportunities.

Requirements are the primary focus in the systems engineering process because the process's primary purpose is to transform the requirements into designs. The process develops these designs within the constraints. They eventually must be verified to meet both the requirements and constraints.

Types of Requirements

Requirements are categorized in several ways. The following are common categorizations of requirements that relate to technical management:

Customer Requirements: Statements of fact and assumptions that define the expectations of the system in terms of mission objectives, environment, constraints, and measures of effectiveness and suitability (MOE/MOS). The customers are those that perform the eight primary functions of systems engineering (Chapter 1), with special emphasis on the operator as the key customer. Operational requirements will define the basic need and, at a minimum, answer the questions posed in Figure 4-1.

Operational distribution or deployment: Where will the system be used?

Mission profile or scenario: How will the system accomplish its mission objective?

Performance and related parameters: What are the critical system parameters to accomplish the mission?

Utilization environments: How are the various system components to be used?

Effectiveness requirements: How effective or efficient must the system be in performing its mission?

Operational life cycle: How long will the system be in use by the user?

Environment: What environments will the system be expected to operate in an effective manner?

Figure 4-1. Operational Requirements – Basic Questions

Functional Requirements: The necessary task, action or activity that must be accomplished. Functional (what has to be done) requirements identified in requirements analysis will be used as the top-level functions for functional analysis.

Performance Requirements: The extent to which a mission or function must be executed; generally measured in terms of quantity, quality, coverage, timeliness or readiness. During requirements analysis, performance (how well does it have to be done) requirements will be interactively developed across all identified functions based on system life cycle factors; and characterized in terms of the degree of certainty in their estimate, the degree of criticality to system success, and their relationship to other requirements.

Design Requirements: The "build to," "code to," and "buy to" requirements for products and "how to execute" requirements for processes expressed in technical data packages and technical manuals.

Derived Requirements: Requirements that are implied or transformed from higher-level requirement. For example, a requirement for long range or high speed may result in a design requirement for low weight.

Allocated Requirements: A requirement that is established by dividing or otherwise allocating a high-level requirement into multiple lower-level requirements. Example: A 100-pound item that consists of two subsystems might result in weight requirements of 70 pounds and 30 pounds for the two lower-level items.

Attributes of Good Requirements

The attributes of good requirements include the following:

- A requirement must be achievable. It must reflect need or objective for which a solution is technically achievable at costs considered affordable.

- It must be verifiable—that is, not defined by words such as excessive, sufficient, resistant, etc. The expected performance and functional utility must be expressed in a manner that allows verification to be objective, preferably quantitative.

- A requirement must be unambiguous. It must have but one possible meaning.

- It must be complete and contain all mission profiles, operational and maintenance concepts, utilization environments and constraints. All information necessary to understand the customer's need must be there.

- It must be expressed in terms of need, not solution; that is, it should address the "why" and "what" of the need, not how to do it.

- It must be consistent with other requirements. Conflicts must be resolved up front.

- It must be appropriate for the level of system hierarchy. It should not be too detailed that it constrains solutions for the current level of design. For example, detailed requirements relating to components would not normally be in a system-level specification.

4.2 REQUIREMENTS ANALYSIS

Requirements analysis involves defining customer needs and objectives in the context of planned customer use, environments, and identified system characteristics to determine requirements for system functions. Prior analyses are reviewed and updated, refining mission and environment definitions to support system definition.

Requirements analysis is conducted iteratively with functional analysis to optimize performance requirements for identified functions, and to verify that synthesized solutions can satisfy customer requirements. The purpose of Requirements Analysis is to:

- Refine customer objectives and requirements;

- Define initial performance objectives and refine them into requirements;

- Identify and define constraints that limit solutions; and

- Define functional and performance requirements based on customer provided measures of effectiveness.

In general, Requirements Analysis should result in a clear understanding of:

- Functions: What the system has to do,

- Performance: How well the functions have to be performed,

- Interfaces: Environment in which the system will perform, and

- Other requirements and constraints.

The understandings that come from requirements analysis establish the basis for the functional and physical designs to follow. Good requirements analysis is fundamental to successful design definition.

Inputs

Typical inputs include customer needs and objectives, missions, MOE/MOS, environments, key performance parameters (KPPs), technology base, output requirements from prior application of SEP, program decision requirements, and suitability requirements. (See Figure 4-2 for additional considerations.)

Input requirements must be comprehensive and defined for both system products and system processes such as development, manufacturing, verification, deployment, operations, support, training and disposal (eight primary functions).

Role of Integrated Teams

The operator customers have expertise in the operational employment of the product or item being developed. The developers (government and contractor) are not necessarily competent in the operational aspects of the system under development. Typically, the operator's need is neither clearly nor completely expressed in a way directly

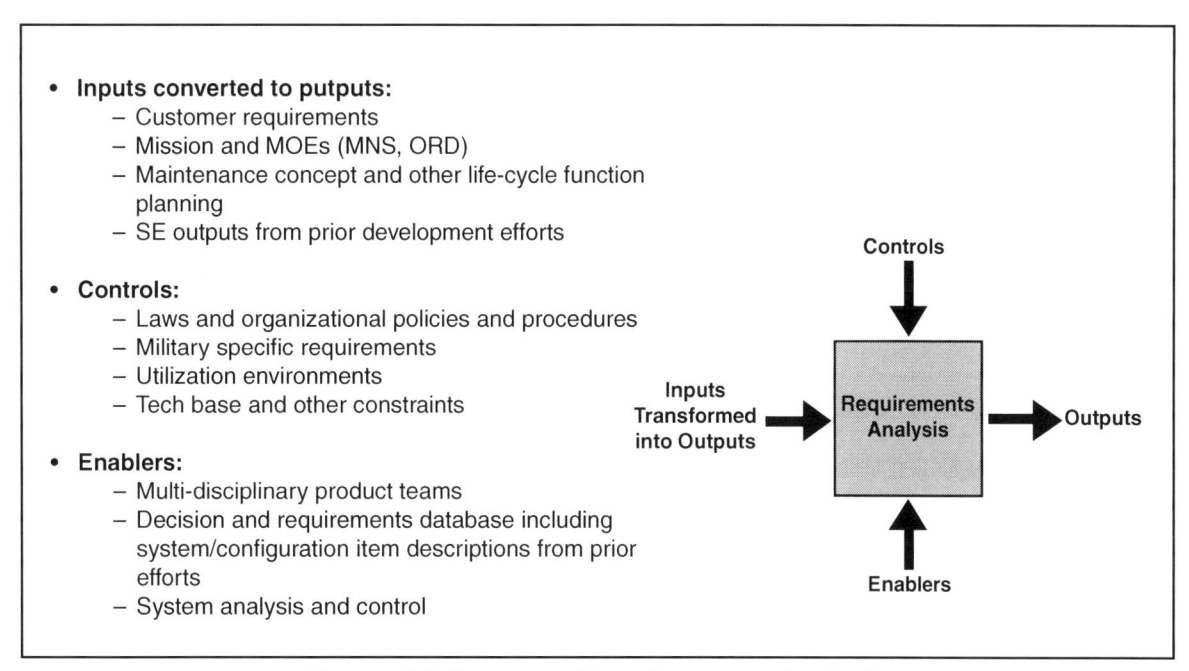

- **Inputs converted to putputs:**
  - Customer requirements
  - Mission and MOEs (MNS, ORD)
  - Maintenance concept and other life-cycle function planning
  - SE outputs from prior development efforts

- **Controls:**
  - Laws and organizational policies and procedures
  - Military specific requirements
  - Utilization environments
  - Tech base and other constraints

- **Enablers:**
  - Multi-disciplinary product teams
  - Decision and requirements database including system/configuration item descriptions from prior efforts
  - System analysis and control

Figure 4-2. Inputs to Requirements Analysis

usable by developers. It is unlikely that developers will receive a well-defined problem from which they can develop the system specification. Thus, teamwork is necessary to understand the problem and to analyze the need. It is imperative that customers are part of the definition team.

On the other hand, customers often find it easier to describe a system that attempts to solve the problem rather than to describe the problem itself. Although these "solutions" may be workable to some extent, the optimum solution is obtained through a proper technical development effort that properly balances the various customer mission objectives, functions, MOE/MOS, and constraints. An integrated approach to product and process development will balance the analysis of requirements by providing understanding and accommodation among the eight primary functions.

Requirements Analysis Questions

Requirements Analysis is a process of inquiry and resolution. The following are typical questions that can initiate the thought process:

- What are the reasons behind the system development?

- What are the customer expectations?

- Who are the users and how do they intend to use the product?

- What do the users expect of the product?

- What is their level of expertise?

- With what environmental characteristics must the system comply?

- What are existing and planned interfaces?

- What functions will the system perform, expressed in customer language?

- What are the constraints (hardware, software, economic, procedural) to which the system must comply?

- What will be the final form of the product: such as model, prototype, or mass production?

This list can start the critical, inquisitive outlook necessary to analyze requirements, but it is only the beginning. A tailored process similar to the one at the end of this chapter must be developed to produce the necessary requirements analysis outputs.

4.3 REQUIREMENTS ANALYSIS OUTPUTS

The requirements that result from requirements analysis are typically expressed from one of three perspectives or views. These have been described as the Operational, Functional, and Physical views. All three are necessary and must be coordinated to fully understand the customers' needs and objectives. All three are documented in the decision database.

Operational View

The Operational View addresses how the system will serve its users. It is useful when establishing requirements of "how well" and "under what condition." Operational view information should be documented in an operational concept document that identifies:

- Operational need definition,

- System mission analysis,

- Operational sequences,

- Operational environments,

- Conditions/events to which a system must respond,

- Operational constraints on system,

- Mission performance requirements,

- User and maintainer roles (defined by job tasks and skill requirements or constraints),

- Structure of the organizations that will operate, support and maintain the system, and

- Operational interfaces with other systems.

Analyzing requirements requires understanding the operational and other life cycle needs and constraints.

Functional View

The Functional View focuses on WHAT the system must do to produce the required operational behavior. It includes required inputs, outputs, states, and transformation rules. The functional requirements, in combination with the physical requirements shown below, are the primary sources of the requirements that will eventually be reflected in the system specification. Functional View information includes:

- System functions,

- System performance,
  - Qualitative — how well
  - Quantitative — how much, capacity
  - Timeliness — how often

- Tasks or actions to be performed,

- Inter-function relationships,

- Hardware and software functional relationships,

- Performance constraints,

- Interface requirements including identification of potential open-system opportunities (potential standards that could promote open systems should be identified),

- Unique hardware or software, and

- Verification requirements (to include inspection, analysis/simulation, demo, and test).

Physical View

The Physical View focuses on HOW the system is constructed. It is key to establishing the physical interfaces among operators and equipment, and technology requirements. Physical View information would normally include:

- Configuration of System:
  - Interface descriptions,
  - Characteristics of information displays and operator controls,
  - Relationships of operators to system/physical equipment, and
  - Operator skills and levels required to perform assigned functions.

- Characterization of Users:
  - Handicaps (special operating environments), and
  - Constraints (movement or visual limitations).

- System Physical Limitations:
  - Physical limitations (capacity, power, size, weight),
  - Technology limitations (range, precision, data rates, frequency, language),
  - Government Furnished Equipment (GFE), Commercial-Off-the-Shelf (COTS), Nondevelopmental Item (NDI), reusability requirements, and
  - Necessary or directed standards.

4.4 SUMMARY POINTS

- An initial statement of a need is seldom defined clearly.

- A significant amount of collaboration between various life cycle customers is necessary to produce an acceptable requirements document.

- Requirements are a statement of the problem to be solved. Unconstrained and nonintegrated requirements are seldom sufficient for designing a solution.

- Because requirements from different customers will conflict, constraints will limit options, and resources are not unlimited; trade studies must be accomplished in order to select a balanced set of requirements that provide feasible solutions to customer needs.

SUPPLEMENT 4-A

A PROCEDURE FOR REQUIREMENTS ANALYSIS

The following section provides a list of tasks that represents a plan to analyze requirements. Part of this notional process is based on the 15 requirements analysis tasks listed in IEEE P1220. This industry standard and others should be consulted when preparing engineering activities to help identify and structure appropriate activities.

As with all techniques, the student should be careful to tailor; that is, add or subtract, as suits the particular system being developed. Additionally, these tasks, though they build on each other, should not be considered purely sequential. Every task contributes understanding that may cause a need to revisit previous task decisions. This is the nature of all System Engineering activities.

Preparation: Establish and Maintain Decision Database

When beginning a systems engineering process, be sure that a system is in place to record and manage the decision database. The decision database is an historical database of technical decisions and requirements for future reference. It is the primary means for maintaining requirements traceability. This database decision management system must be developed or the existing system must be reviewed and upgraded as necessary to accommodate the new stage of product development. A key part of this database management system is a Requirements Traceability Matrix that maps requirements to subsystems, configuration items, and functional areas.

This must be developed, updated, and reissued on a regular basis. All requirements must be recorded. *Remember: If it is not recorded, it cannot be an approved requirement!*

The 15 Tasks of IEEE P1220

The IEEE Systems Engineering Standard offers a process for performing Requirements Analysis that comprehensively identifies the important tasks that must be performed. These 15 task areas to be analyzed follow and are shown in Figure 4-3.

---

1. Customer expectations
2. Project and enterprise constraints
3. External constraints
4. Operational scenarios
5. Measure of effectiveness (MOEs)
6. System boundaries
7. Interfaces
8. Utilization environments
9. LIfe cycle
10. Functional requirements
11. Performance requirements
12. Modes of operation
13. Technical performance measures
14. Physical characteristics
15. Human systems integration

Figure 4-3. IEEE P1220 Requirements Analysis Task Areas

Task 1. Customer Expectations

Define and quantify customer expectations. They may come from any of the eight primary functions, operational requirements documents, mission needs, technology-based opportunity, direct communications with customer, or requirements from a higher system level. The purpose of this task is to determine what the customer wants the system to accomplish, and how well each function must be accomplished. This should include natural and induced environments in which the product(s) of the system must operate or be used, and constraints (e.g. funding, cost, or price objectives, schedule, technology, nondevelopmental and reusable items, physical characteristics, hours of operation per day, on-off sequences, etc.).

Task 2. Project and Enterprise Constraints

Identify and define constraints impacting design solutions. Project specific constraints can include:

- Approved specifications and baselines developed from prior applications of the Systems Engineering Process,
- Costs,
- Updated technical and project plans,
- Team assignments and structure,
- Control mechanisms, and
- Required metrics for measuring progress.

Enterprise constraints can include:

- Management decisions from a preceding technical review,
- Enterprise general specifications,
- Standards or guidelines,
- Policies and procedures,
- Domain technologies, and
- Physical, financial, and human resource allocations to the project.

Task 3. External Constraints

Identify and define external constraints impacting design solutions or implementation of the Systems Engineering Process activities. External constraints can include:

- Public and international laws and regulations,
- Technology base,
- Compliance requirements: industry, international, and other general specifications, standards, and guidelines which require compliance for legal, interoperability, or other reasons,
- Threat system capabilities, and
- Capabilities of interfacing systems.

Task 4. Operational Scenarios

Identify and define operational scenarios that scope the anticipated uses of system product(s). For each operational scenario, define expected:

- Interactions with the environment and other systems, and
- Physical interconnectivities with interfacing systems, platforms, or products.

Task 5. Measures of Effectiveness and Suitability (MOE/MOS)

Identify and define systems effectiveness measures that reflect overall customer expectations and satisfaction. MOEs are related to how well the system must perform the customer's mission. Key MOEs include mission performance, safety, operability, reliability, etc. MOSs are related to how well the system performs in its intended environment and includes measures of supportability, maintainability, ease of use, etc.

Task 6. System Boundaries

Define system boundaries including:

- Which system elements are under design control of the performing activity and which fall outside of their control, and

- The expected interactions among system elements under design control and external and/or higher-level and interacting systems outside the system boundary (including open systems approaches).

Task 7. Interfaces

Define the functional and physical interfaces to external or higher-level and interacting systems, platforms, and/or products in quantitative terms (include open systems approach). Functional and physical interfaces would include mechanical, electrical, thermal, data, control, procedural, and other interactions. Interfaces may also be considered from an internal/external perspective. Internal interfaces are those that address elements inside the boundaries established for the system addressed. These interfaces are generally identified and controlled by the contractor responsible for developing the system. External interfaces, on the other hand, are those which involve entity relationships outside the established boundaries, and these are typically defined and controlled by the government.

Task 8. Utilization Environments

Define the environments for each operational scenario. All environmental factors (natural or induced) which may impact system performance must be identified and defined. Environmental factors include:

- Weather conditions (e.g., rain, snow, sun, wind, ice, dust, fog),

- Temperature ranges,

- Topologies (e.g., ocean, mountains, deserts, plains, vegetation),

- Biological (e.g., animal, insects, birds, fungi),

- Time (e.g., dawn, day, night, dusk), and

- Induced (e.g., vibration, electromagnetic, chemical).

Task 9. Life Cycle Process Concepts

Analyze the outputs of tasks 1-8 to define key life cycle process requirements necessary to develop, produce, test, distribute, operate, support, train, and dispose of system products under development. Use integrated teams representing the eight primary functions. Focus should be on the cost drivers and higher risk elements that are anticipated to impact supportability and affordability over the useful life of the system.

Task 10. Functional Requirements

Define what the system must accomplish or must be able to do. Functions identified through requirements analysis will be further decomposed during functional analysis and allocation.

Task 11. Performance Requirements

Define the performance requirements for each higher-level function performed by the system. Primary focus should be placed on performance requirements that address the MOEs, and other KPPs established in test plans or identified as interest items by oversight authorities.

Task 12. Modes of Operation

Define the various modes of operation for the system products under development. Conditions (e.g., environmental, configuration, operational, etc.) that determine the modes of operation should be included in this definition.

Task 13. Technical Performance Measures (TPMs)

Identify the key indicators of system performance that will be tracked during the design process. Selection of TPMs should be limited to critical technical thresholds and goals that, if not met, put the project at cost, schedule, or performance risk. TPMs involve tracking the actual versus planned progress of KPPs such that the manager can make judgments about technical progress on a by-exception basis. To some extent TPM selection is phase dependent. They must be reconsidered at each systems engineering process step and at the beginning of each phase.

Task 14. Physical Characteristics

Identify and define required physical characteristics (e.g., color, texture, size, weight, buoyancy) for the system products under development. Identify which physical characteristics are true constraints and which can be changed, based on trade studies.

Task 15. Human Factors

Identify and define human factor considerations (e.g., physical space limits, climatic limits, eye movement, reach, ergonomics) which will affect operation of the system products under development. Identify which human systems integration are constraints and which can be changed based on trade studies.

Follow-on Tasks

The follow-on tasks are related to the iterative nature of the Systems Engineering Process:

Integrate Requirements:

Take an integrated team approach to requirements determination so that conflicts among and between requirements are resolved in ways that result in design requirements that are balanced in terms of both risk and affordability.

Validate Requirements:

During Functional Analysis and Allocation, validate that the derived functional and performance can be traced to the operational requirements.

Verify Requirements:

- Coordinate design, manufacturing, deployment and test processes,

- Ensure that requirements are achievable and testable,

- Verify that the design-to-cost goals are achievable, and

- Verify that the functional and physical architectures defined during Functional Analysis/Allocation and Synthesis meet the integrated technical, cost, and schedule requirements within acceptable levels of risk.

CHAPTER 5

FUNCTIONAL ANALYSIS AND ALLOCATION

5.1 INTRODUCTION

The purpose of this systems engineering process activity is to transform the functional, performance, interface and other requirements that were identified through requirements analysis into a coherent description of system functions that can be used to guide the Design Synthesis activity that follows. The designer will need to know what the system must do, how well, and what constraints will limit design flexibility.

This is accomplished by arranging functions in logical sequences, decomposing higher-level functions into lower-level functions, and allocating performance from higher- to lower-level functions. The tools used include functional flow block diagrams and timeline analysis; and the product is a functional architecture, i.e., a description of the system—but in terms of functions and performance parameters, rather than a physical description. Functional Analysis and Allocation facilitates traceability from requirements to the solution descriptions that are the outcome of Design Synthesis.

Functions are discrete actions (use action verbs) necessary to achieve the system's objectives. These functions may be stated explicitly, or they may be derived from stated requirements. The functions will ultimately be performed or accomplished through use of equipment, personnel, facilities, software, or a combination.

5.2 FUNCTIONAL ANALYSIS AND ALLOCATION

Functional and performance requirements at any level in the system are developed from higher-level requirements. Functional Analysis and Allocation is repeated to define successively lower-level functional and performance requirements, thus defining architectures at ever-increasing levels of detail. System requirements are allocated and defined in sufficient detail to provide design and verification criteria to support the integrated system design.

This top-down process of translating system-level requirements into detailed functional and performance design criteria includes:

- Defining the system in functional terms, then decomposing the top-level functions into subfunctions. That is, identifying at successively lower levels what actions the system has to do,

- Translating higher-level performance requirements into detailed functional and performance design criteria or constraints. That is, identifying how well the functions have to be performed,

- Identifying and defining all internal and external functional interfaces,

- Identifying functional groupings to minimize and control interfaces (functional partitioning),

- Determining the functional characteristics of existing or directed components in the system and incorporating them in the analysis and allocation,

- Examining all life cycle functions, including the eight primary functions, as appropriate for the specific project,

- Performing trade studies to determine alternative functional approaches to meet requirements, and

- Revisiting the requirements analysis step as necessary to resolve functional issues.

The objective is to identify the functional, performance, and interface design requirements; it is not to design a solution…yet!

Functional Partitioning

Functional partitioning is the process of grouping functions that logically fit with the components likely to be used, and to minimize functional interfaces. Partitioning is performed as part of functional decomposition. It identifies logical groupings of functions that facilitate the use of modular components and open-system designs. Functional partitioning is also useful in understanding how existing equipment or components (including commercial) will function with or within the system.

Requirements Loop

During the performance of the Functional Analysis and Allocation process, it is expected that revisiting the requirements analysis process will be necessary. This is caused by the emergence of functional issues that will require re-examination of the higher-level requirements. Such issues might include directed components or standards that cause functional conflict, identification of a revised approach to functional sequencing, or, most likely, a conflict caused by mutually incompatible requirements.

Figure 5-1 gives an overview of the basic parameters of Functional Analysis and Allocation. The output of the process is the functional architecture. In its most basic form, the functional architecture is a simple hierarchical decomposition of the functions with associated performance requirements. As the architecture definition is refined and made more specific with the performance of the

- **Outputs:**
    - Functional architecture and supporting detail

- **Inputs:**
    - Outputs of the Requirements Analysis

- **Enablers:**
    - Multi-discipline product teams, decision database; Tools & Models, such as QFD, Functional Flow Block Diagrams, IDEF, N2 charts, Requirement Allocation Sheet, Timelines, Data Flow Diagrams, State/Mode Diagrams, Behavior Diagrams

- **Controls:**
    - Constraints; GFE, COTS, & Reusable S/W; System concept & subsystem choices; organizational procedures

- **Activities:**
    - Define system states and modes
    - Define system functions & external interfaces
    - Define functional interfaces
    - Allocate performance requirements to functions
    - Analyze performance
    - Analyze timing and resources
    - Analyze failure mode effects and criticality
    - Define fault detection and recovery behavior
    - Integrate functions

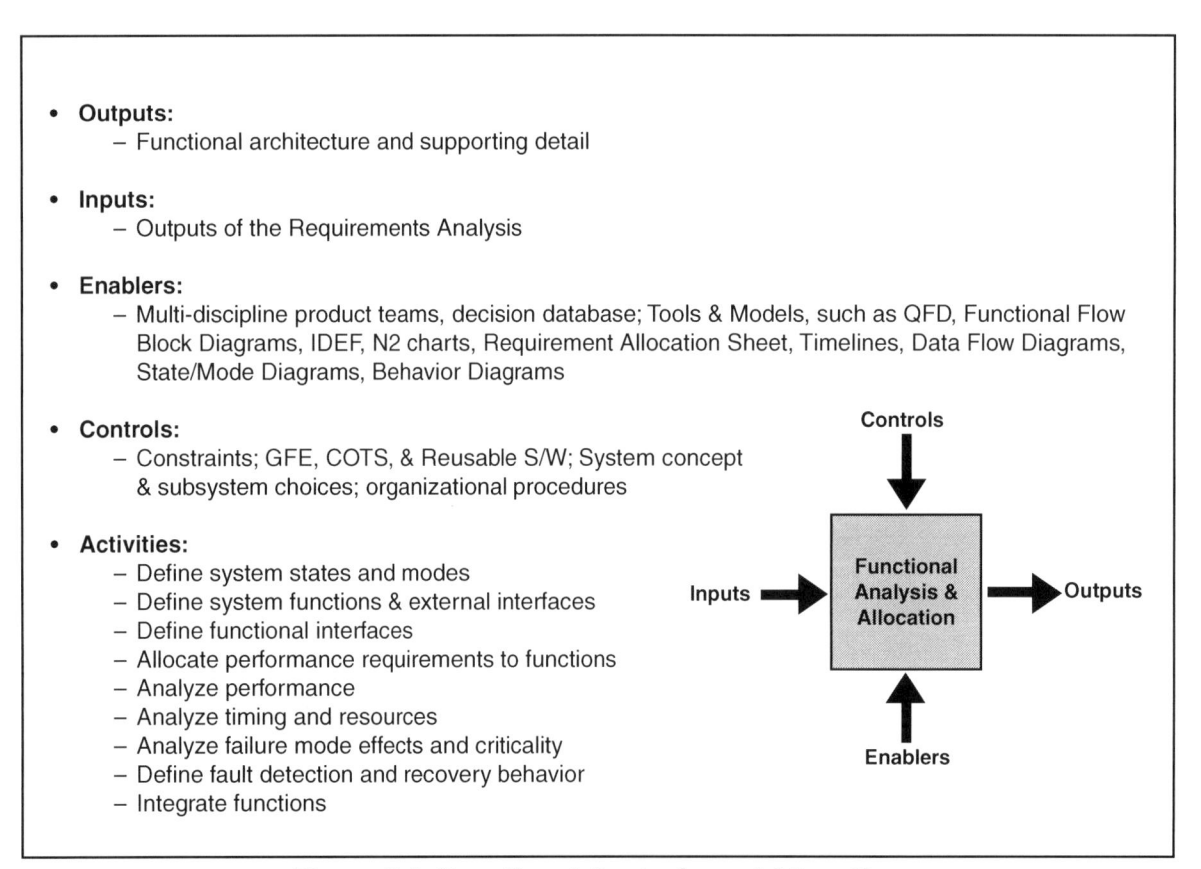

Figure 5-1. Functional Analysis and Allocation

activities listed in Figure 5-1, the functional architecture becomes more detailed and comprehensive. These activities provide a functional architecture with sufficient detail to support the Design Synthesis. They are performed with the aid of traditional tools that structure the effort and provide documentation for traceability. There are many tools available. The following are traditional tools that represent and explain the primary tasks of Functional Analysis and Allocation (several of these are defined and illustrated beginning on page 49):

- Functional flow block diagrams that define task sequences and relationships,

- IDEF0 diagrams that define process and data flows,

- Timeline analyses that define the time sequence of time critical functions, and

- Requirements allocation sheets that identify allocated performance and establish traceability of performance requirements.

5.3 FUNCTIONAL ARCHITECTURE

The functional architecture is a top-down decomposition of system functional and performance requirements. The architecture will show not only the functions that have to be performed, but also the logical sequencing of the functions and performance requirements associated with the functions. It also includes the functional description of existing and government-furnished items to be used in the system. This may require reverse engineering of these existing components.

The functional architecture produced by the Functional Analysis and Allocation process is the detailed package of documentation developed to analyze the functions and allocate performance requirements. It includes the functional flow block diagrams, timeline sheets, requirements allocation sheets, IDEF0 diagrams, and all other documentation developed to describe the functional characteristics of the system. However, there is a basic logic to the functional architecture, which in its preliminary form is presented in the example of Figure 5-2. The Functional Analysis and Allocation process would normally begin with the

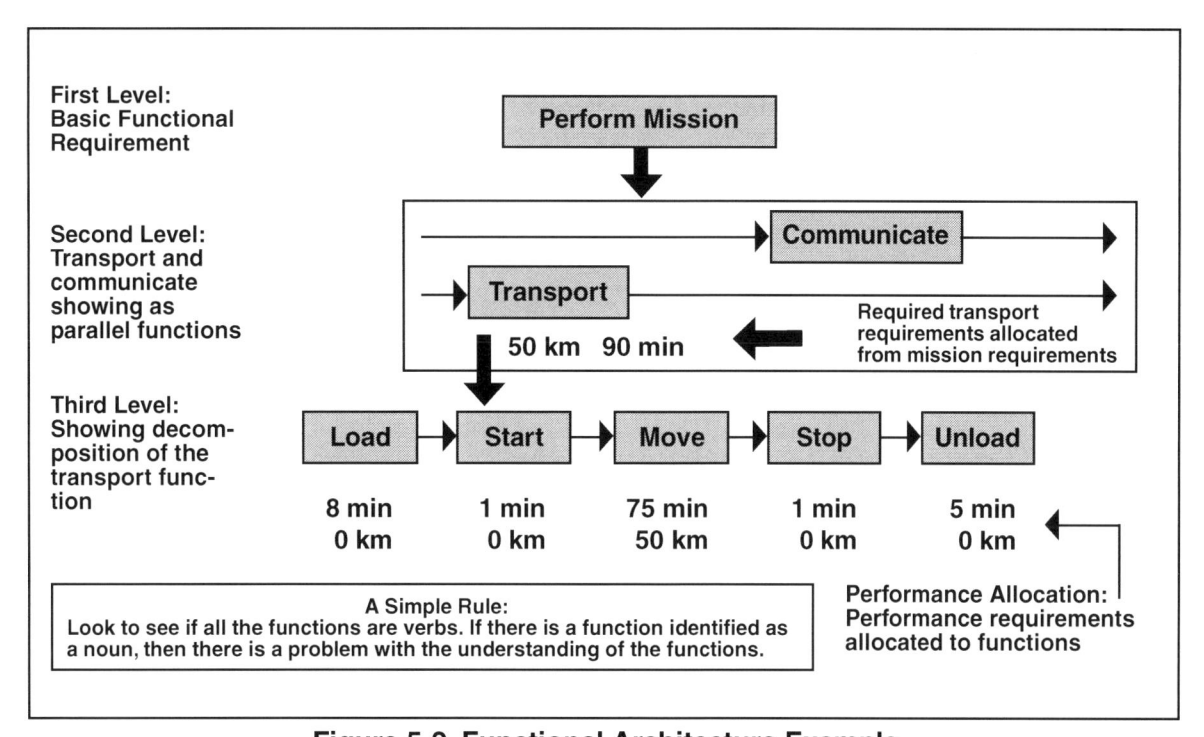

Figure 5-2. Functional Architecture Example

IPT drafting such a basic version of the architecture. This would generally give the IPT an understanding of the scope and direction of the effort.

Functional Architecture Example

The Marine Corps has a requirement to transport troops in squad-level units over a distance of 50 kilometers. Troops must be transported within 90 minutes from the time of arrival of the transport system. Constant communication is required during the transportation of troops. Figure 5-2 illustrates a preliminary functional architecture for this simple requirement.

5.4 SUMMARY POINTS

Functional analysis begins with the output of requirements analysis (that is, the identification of higher-level functional and performance requirements). Functional Analysis and Allocation consists of decomposition of higher-level functions to lower-levels and then allocation of requirements to those functions.

- There are many tools available to support the development of a Functional Architecture, such as: functional-flow block diagrams, timeline analysis sheet, requirements allocation sheet, Integrated Definition, and others.

- Use of the tools illustrated in this chapter is not mandatory, but the process they represent is:

  – Define task sequences and relationships (functional flow block diagram (FFBD)),

  – Define process and data flows (IDEF0 diagrams),

  – Define the time sequence of time-critical functions (timeline analysis sheets (TLS)), and

  – Allocate performance and establish traceability of performance requirements (requirements allocation sheets (RAS)).

SUPPLEMENT 5-A

FUNCTIONAL FLOW BLOCK DIAGRAM

The purpose of the functional flow block diagram (FFBD) is to describe system requirements in functional terms.

Objectives

The FFBD is structured to ensure that:

- All life cycle functions are covered.

- All elements of system are identified and defined (e.g. prime equipment, training, spare parts, data, software, etc.).

- System support requirements are identified to specific system functions.

- Proper sequencing of activities and design relationships are established including critical design interfaces.

Characteristics

The FFBD is functionally oriented—not solution oriented. The process of defining lower-level functions and sequencing relationships is often referred to as functional decomposition. It allows traceability vertically through the levels. It is a key step in developing the functional architecture from which designs may be synthesized.

Figure 5-3 shows the flow-down structure of a set of FFBDs and Figure 5-4 shows the format of an FFBD.

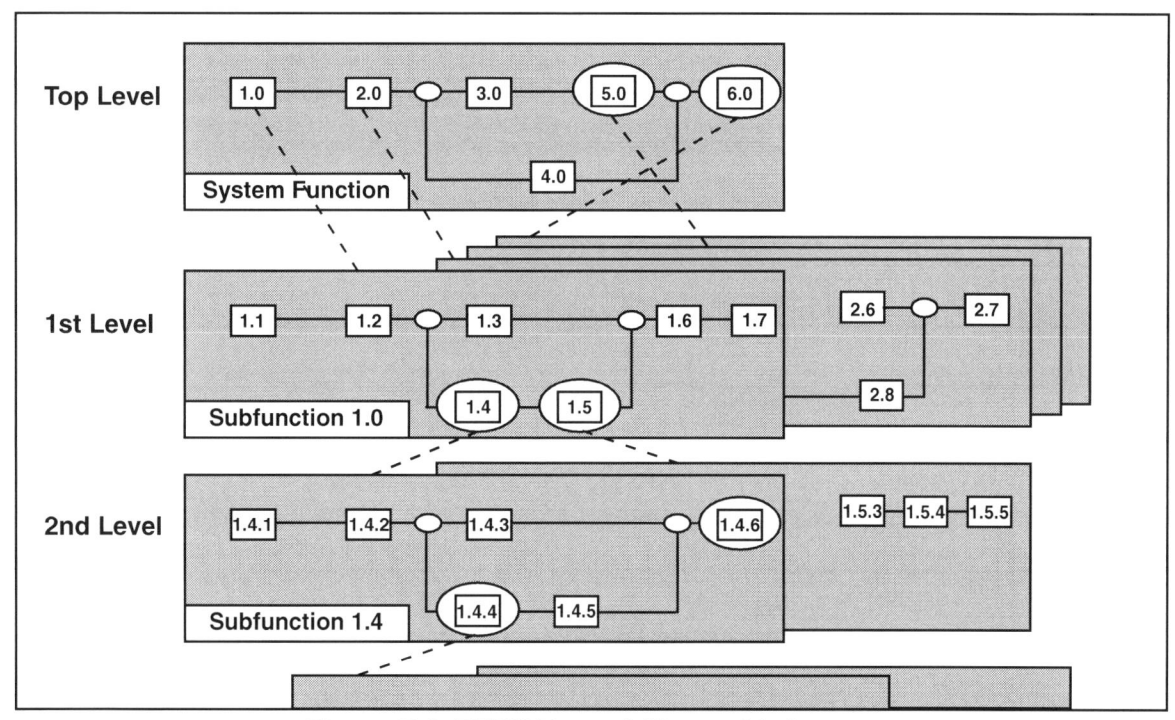

Figure 5-3. FFBD Traceability and Indenture

Key FFBD Attributes

Function block: Each function on an FFBD should be separate and be represented by single box (solid line). Each function needs to stand for definite, finite, discrete action to be accomplished by system elements.

Function numbering: Each level should have a consistent number scheme and provide information concerning function origin. (E.g., top level— 1.0, 2.0, 3.0, etc; first indenture (level 2)—1.1, 1.2, 1.3, etc; second indenture (level 3)—1.1.1, 1.1.2, 1.1.3, etc.) These numbers establish identification and relationships that will carry through all Functional Analysis and Allocation activities and facilitate traceability from lower to top levels.

Functional reference: Each diagram should contain a reference to other functional diagrams by using a functional reference (box in brackets).

Flow connection: Lines connecting functions should only indicate function flow and not a lapse in time or intermediate activity.

Flow direction: Diagrams should be laid out so that the flow direction is generally from left to right. Arrows are often used to indicate functional flows.

Summing gates: A circle is used to denote a summing gate and is used when AND/OR is present. AND is used to indicate parallel functions and all conditions must be satisfied to proceed. OR is used to indicate that alternative paths can be satisfied to proceed.

GO and NO-GO paths: "G" and "bar G" are used to denote "go" and "no-go" conditions. These symbols are placed adjacent to lines leaving a particular function to indicate alternative paths.

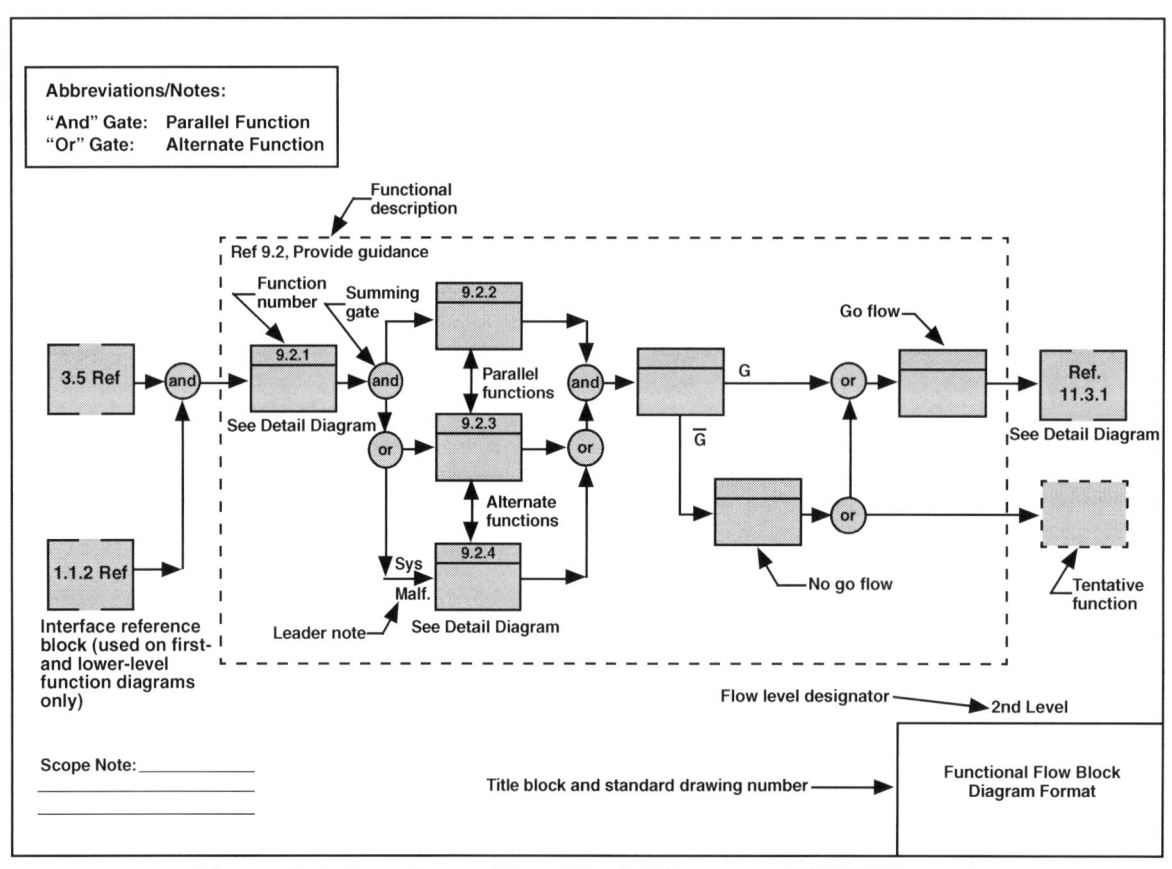

Figure 5-4. Functional Flow Block Diagrams (FFBD) Format

SUPPLEMENT 5-B

IDEF0

Integration Definition for Function Modeling (IDEF0) is a common modeling technique for the analysis, development, re-engineering, and integration of information systems; business processes; or software engineering analysis. Where the FFBD is used to show the functional flow of a product, IDEF0 is used to show data flow, system control, and the functional flow of life cycle processes.

IDEF0 is capable of graphically representing a wide variety of business, manufacturing and other types of enterprise operations to any level of detail. It provides rigorous and precise description, and promotes consistency of usage and interpretation. It is well-tested and proven through many years of use by government and private industry. It can be generated by a variety of computer graphics tools. Numerous commercial products specifically support development and analysis of IDEF0 diagrams and models.

IDEF0 is a model that consists of a hierarchical series of diagrams, text, and glossary cross-referenced to each other. The two primary modeling components are: functions (represented on a diagram by boxes), and data and objects that interrelate those functions (represented by arrows). As shown by Figure 5-5 the position at which the arrow attaches to a box conveys the specific role of the interface. The controls enter the top of the box. The inputs, the data or objects acted upon by the operation, enter the box from the left. The outputs of the operation leave the right-hand side of the box. Mechanism arrows that provide supporting means for performing the function join (point up to) the bottom of the box.

The IDEF0 process starts with the identification of the prime function to be decomposed. This function is identified on a "Top Level Context Diagram," that defines the scope of the particular IDEF0 analysis. An example of a Top Level Context Diagram for an information system management process is shown in Figure 5-6. From this diagram lower-level diagrams are generated. An example of a derived diagram, called a "child" in

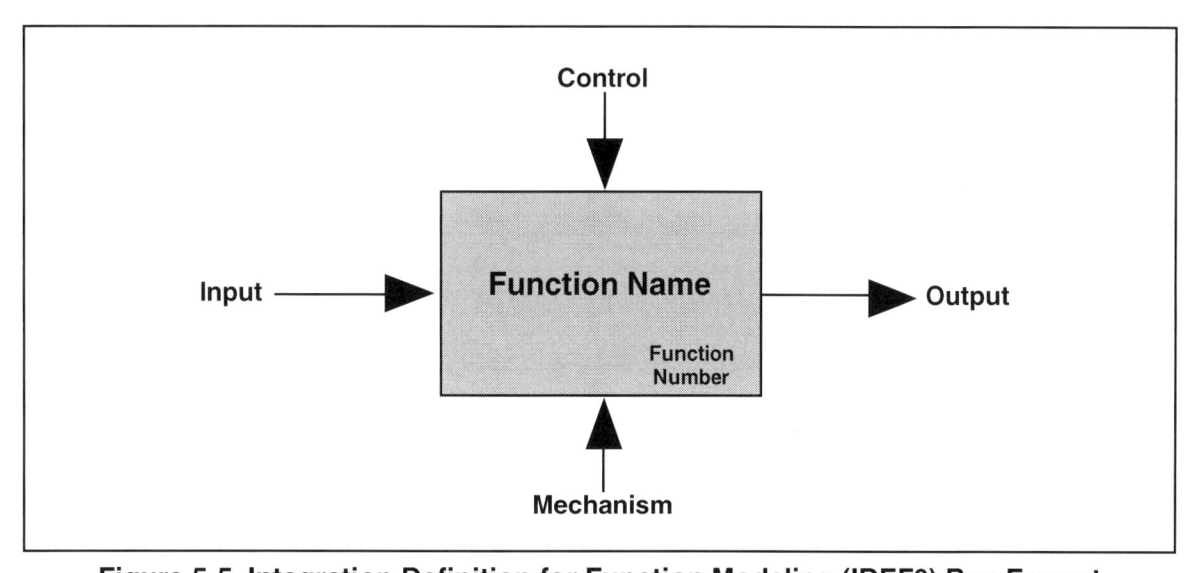

Figure 5-5. Integration Definition for Function Modeling (IDEF0) Box Format

IDEF0 terminology, for a life cycle function is shown in Figure 5-7.

An associated technique, Integration Definition for Information Modeling (IDEF1x), is used to supplement IDEF0 for data intensive systems. The IDEF0 standard, Federal Information Processing Standards Publication 183 (FIPS 183), and the IDEF1x standard (FIPS 184) are maintained by the National Institute of Standards and Technology (NIST).

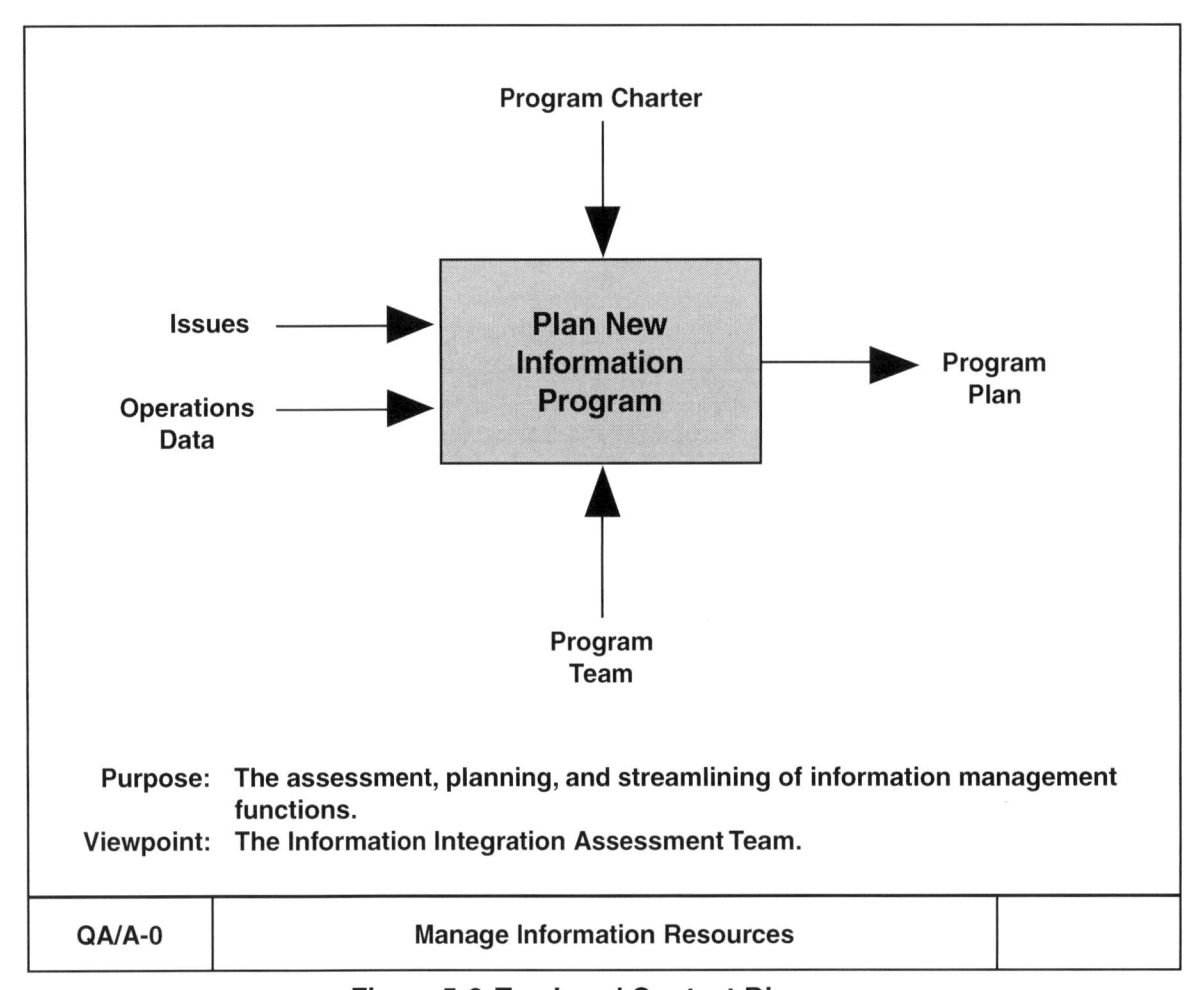

Figure 5-6. Top-Level Context Diagram

Chapter 5 — Functional Analysis and Allocation

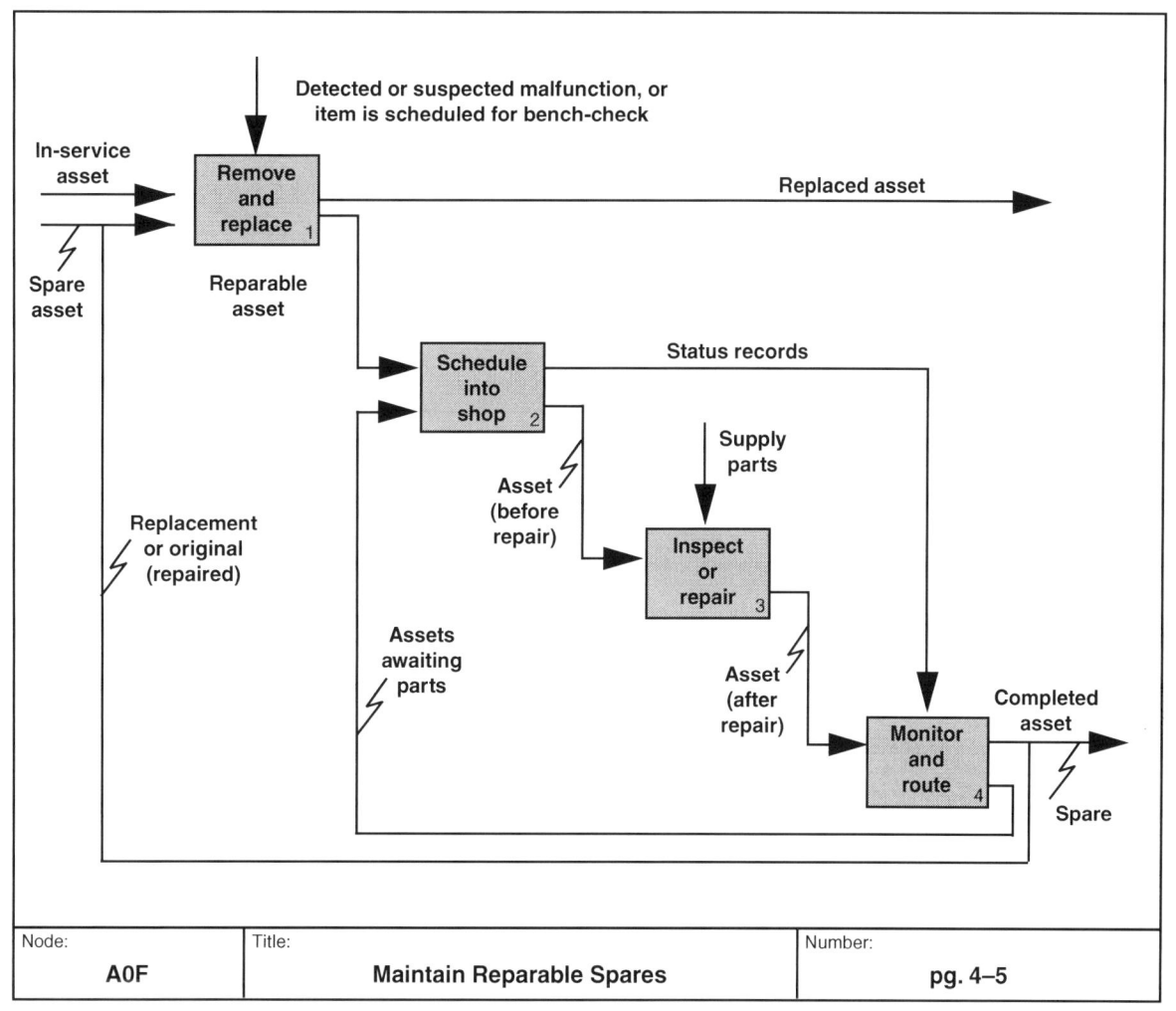

Figure 5-7. IDEF0 Diagram Example

SUPPLEMENT 5-C

TIMELINE ANALYSIS SHEETS

The timeline analysis sheet (TLS) adds detail to defining durations of various functions. It defines concurrency, overlapping, and sequential relationships of functions and tasks. It identifies time critical functions that directly affect system availability, operating time, and maintenance downtime. It is used to identify specific time-related design requirements.

The TLS includes purpose of function and the detailed performance characteristics, criticality of function, and design constraints. It identifies both quantitative and qualitative performance requirements. Initial resource requirements are identified.

Figure 5-8 shows an example of a TLS. The time required to perform function 3.1 and its subfunctions are presented on a bar chart showing how the timelines relate. (Function numbers match the FFBD.)

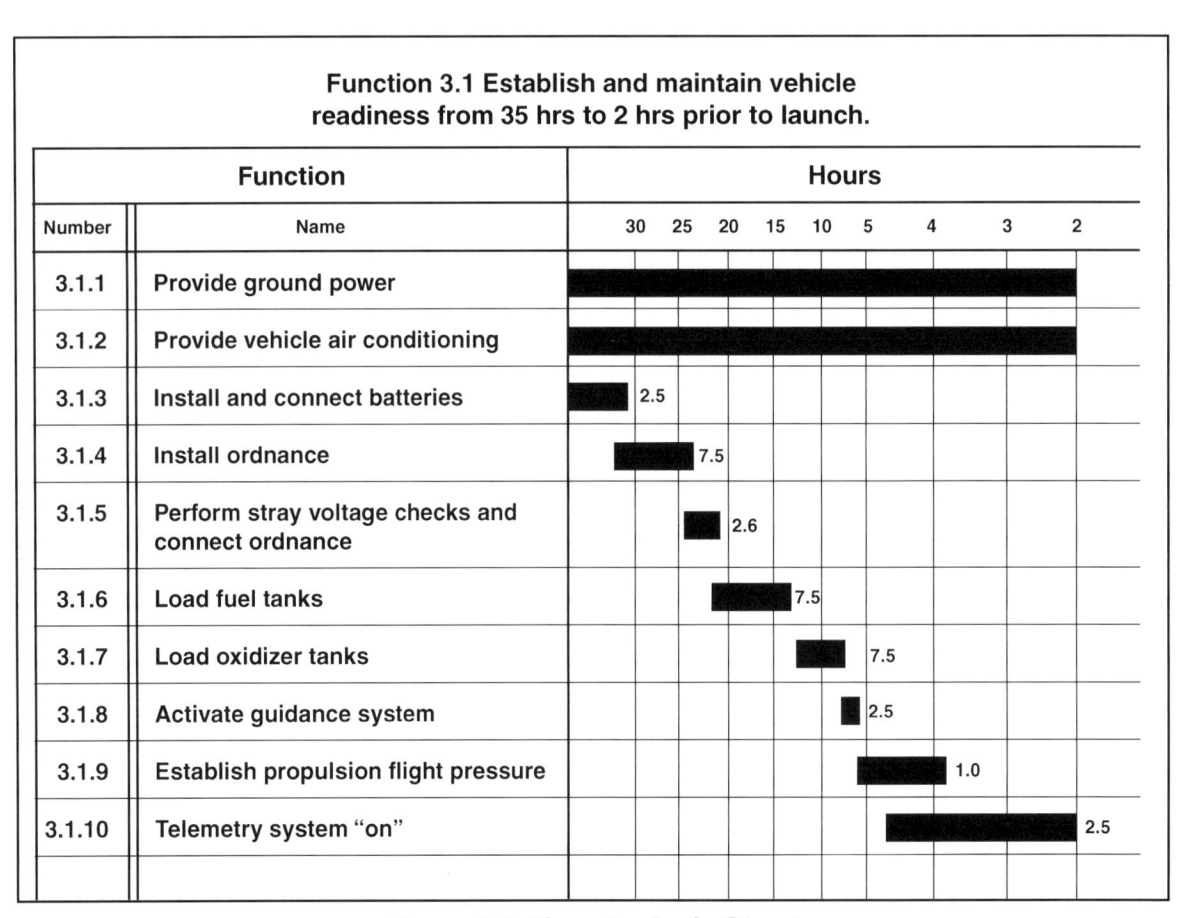

Function 3.1 Establish and maintain vehicle readiness from 35 hrs to 2 hrs prior to launch.

Function Number	Name	Hours
3.1.1	Provide ground power	(30 to 2)
3.1.2	Provide vehicle air conditioning	(30 to 2)
3.1.3	Install and connect batteries	2.5
3.1.4	Install ordnance	7.5
3.1.5	Perform stray voltage checks and connect ordnance	2.6
3.1.6	Load fuel tanks	7.5
3.1.7	Load oxidizer tanks	7.5
3.1.8	Activate guidance system	2.5
3.1.9	Establish propulsion flight pressure	1.0
3.1.10	Telemetry system "on"	2.5

Figure 5-8. Time Analysis Sheet

SUPPLEMENT 5-D

REQUIREMENTS ALLOCATION SHEET

The Requirements Allocation Sheet documents the connection between allocated functions, allocated performance and the physical system. It provides traceability between Functional Analysis and Allocation and Design Synthesis, and shows any disconnects. It is a major tool in maintaining consistency between functional architectures and designs that are based on them. (Function numbers match the FFBD.)

Requirements Allocation Sheet	Functional Flow Diagram Title and No. 2.58.4 Provide Guidance Compartment Cooling	Equipment Identification		
Function Name and No.	Functional Performance and Design Requirements	Facility Rqmnts	Nomen-clature	CI or Detail Spec No.
2.58.4 Provide Guidance Compartment Cooling	The temperature in the guidance compartment must be maintained at the initial calibration temperature of +0.2 Deg F. The initial calibration temperature of the compartment will be between 66.5 and 68.5 Deg F.			
2.58.4.1 Provide Chilled Coolant (Primary)	A storage capacity for 65 gal of chilled liquid coolant (deionized water) is required. The temperature of the stored coolant must be monitored continuously. The stored coolant must be maintained within a temperature range of 40–50 Deg F. for an indefinite period of time. The coolant supplied must be free of obstructive particles 0.5 micron at all times.			

Figure 5-9. Requirements Allocation Sheet (Example)

CHAPTER 6

DESIGN SYNTHESIS

6.1 DESIGN DEVELOPMENT

Design Synthesis is the process by which concepts or designs are developed based on the functional descriptions that are the products of Functional Analysis and Allocation. Design synthesis is a creative activity that develops a physical architecture (a set of product, system, and/or software elements) capable of performing the required functions within the limits of the performance parameters prescribed. Since there may be several hardware and/or software architectures developed to satisfy a given set of functional and performance requirements, synthesis sets the stage for trade studies to select the best among the candidate architectures. The objective of design synthesis is to combine and restructure hardware and software components in such a way as to achieve a design solution capable of satisfying the stated requirements. During concept development, synthesis produces system concepts and establishes basic relationships among the subsystems. During preliminary and detailed design, subsystem and component descriptions are elaborated, and detailed interfaces between all system components are defined.

The physical architecture forms the basis for design definition documentation, such as, specifications, baselines, and work breakdown structures (WBS). Figure 6-1 gives an overview of the basic parameters of the synthesis process.

- **Outputs:**
    - Physical Architecture (Product Elements and Software Code)
    - Decision Database

- **Inputs:**
    - Functional Architecture

- **Enablers:**
    - IPTs, Decision Database, Automated Tools, Models

- **Controls:**
    - Constraints; GFE, COTS, & Reusable S/W; System concept & subsystem choices; organizational procedures

- **Activities:**
    - Allocate functions and constraints to system elements
    - Synthesize system element alternatives
    - Assess technology alternatives
    - Define physical interfaces
    - Define system product WBS
    - Develop life cycle techniques and procedures
    - Integrate system elements
    - Select preferred concept/design

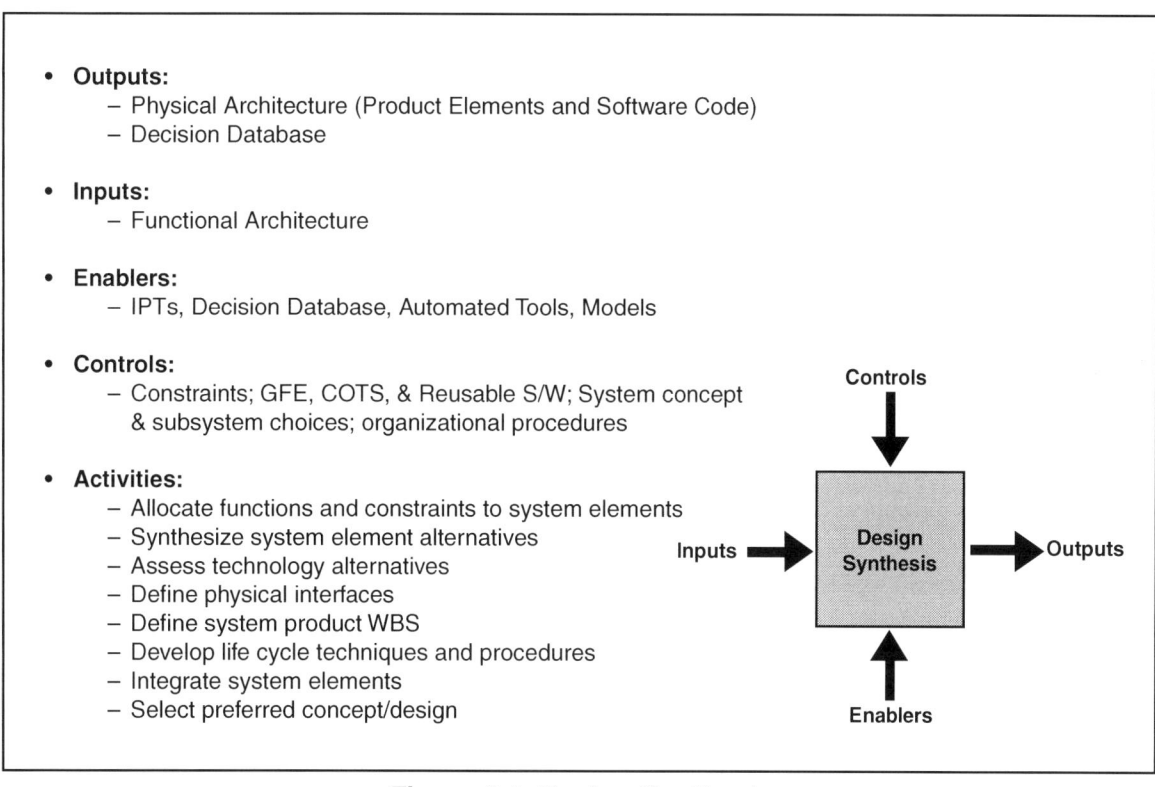

Figure 6-1. Design Synthesis

Characteristics

Physical architecture is a traditional term. Despite the name, it includes software elements as well as hardware elements. Among the characteristics of the physical architecture (the primary output of Design Synthesis) are the following:

- The correlation with functional analysis requires that each physical or software component meets at least one (or part of one) functional requirement, though any component can meet more than one requirement,

- The architecture is justified by trade studies and effectiveness analyses,

- A product WBS is developed from the physical architecture,

- Metrics are developed to track progress among KPPs, and

- All supporting information is documented in a database.

Modular Designs

Modular designs are formed by grouping components that perform a single independent function or single logical task; have single entry and exit points; and are separately testable. Grouping related functions facilitates the search for modular design solutions and furthermore increases the possibility that open-systems approaches can be used in the product architecture.

Desirable attributes of the modular units include low coupling, high cohesion, and low connectivity. Coupling between modules is a measure of their interdependence, or the amount of information shared between two modules. Decoupling modules eases development risks and makes later modifications easier to implement. Cohesion (also called binding) is the similarity of tasks performed within the module. High cohesion is desirable because it allows for use of identical or like (family or series) components, or for use of a single component to perform multiple functions. Connectivity refers to the relationship of internal elements within one module to internal elements within another module. High connectivity is undesirable in that it creates complex interfaces that may impede design, development, and testing.

Design Loop

The design loop involves revisiting the functional architecture to verify that the physical architecture developed is consistent with the functional and performance requirements. It is a mapping between the functional and physical architectures. Figure 6-2 shows an example of a simple physical architecture and how it relates to the functional architecture. During design synthesis, re-evaluation of the functional analysis may be caused by the discovery of design issues that require re-examination of the initial decomposition, performance allocation, or even the higher-level requirements. These issues might include identification of a promising physical solution or open-system opportunities that have different functional characteristics than those foreseen by the initial functional architecture requirements.

6.2 SYNTHESIS TOOLS

During synthesis, various analytical, engineering, and modeling tools are used to support and document the design effort. Analytical devices such as trade studies support decisions to optimize physical solutions. Requirements Allocation Sheets (RAS) provide traceability to the functional and performance requirements. Simple descriptions like the Concept Decription Sheet (CDS) help visualize and communicate the system concept. Logic models, such as the Schematic Block Diagram (SBD), establish the design and the interrelationships within the system.

Automated engineering management tools such as Computer-Aided Design (CAD), Computer-Aided-Systems Engineering (CASE), and the Computer-Aided-Engineering (CAE) can help organize, coordinate and document the design effort. CAD generates detailed documentation describing the product design including SBDs, detailed

	Aircraft				
	Air Frame	Engine	Communications	Nav System	Fire Control
Function Performed					
Preflight check	X	X	X	X	X
Fly					
Load	X				
Taxi	X	X	X		
Take-off	X	X			
Cruise	X	X	X	X	
Recon	X	X	X	X	
Communicate			X		
–					
–					
Surveillance					
–					
–					

← — — PHYSICAL ARCHITECTURE — — →
FUNCTIONAL ARCHITECTURE ↕

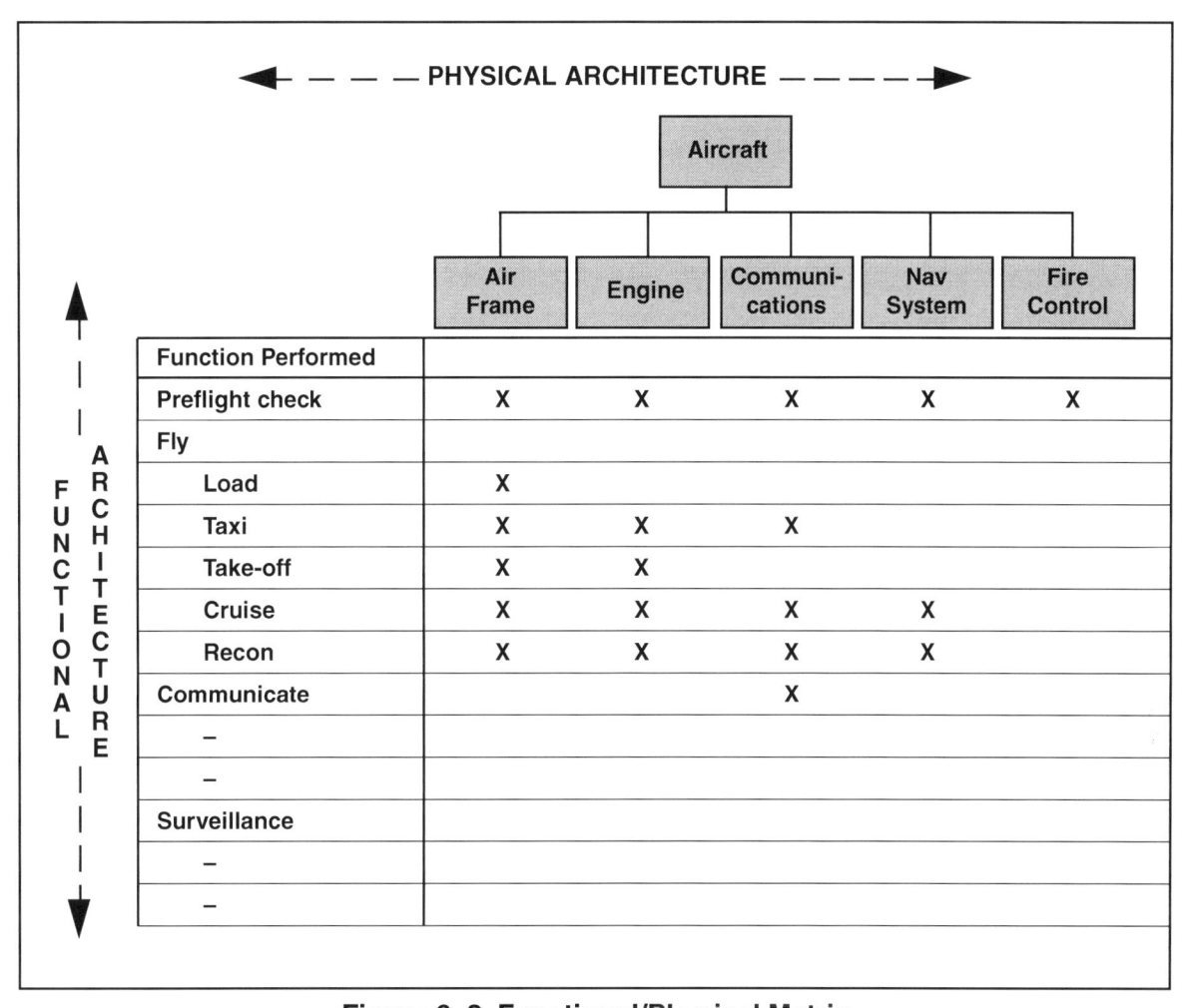

Figure 6–2. Functional/Physical Matrix

drawings, three dimensional and solid drawings, and it tracks some technical performance measurements. CAD can provide significant input for virtual modeling and simulations. It also provides a common design database for integrated design developments. Computer-Aided Engineering can provide system requirements and performance analysis in support of trade studies, analysis related to the eight primary functions, and cost analyses. Computer-Aided Systems Engineering can provide automation of technical management analyses and documentation.

Modeling

Modeling techniques allow the physical product to be visualized and evaluated prior to design decisions. Models allow optimization of hardware and software parameters, permit performance predictions to be made, allow operational sequences to be derived, and permit optimum allocation of functional and performance requirements among the system elements. The traditional logical prototyping used in Design Synthesis is the Schematic Block Diagram.

6.3 SUMMARY POINTS

- Synthesis begins with the output of Functional Analysis and Allocation (the functional architecture). The functional architecture is transformed into a physical architecture by defining physical components needed to perform the functions identified in Functional Analysis and Allocation.

- Many tools are available to support the development of a physical architecture:

    - Define and depict the system concept (CDS),

    - Define and depict components and their relationships (SBD), and

    - Establish traceability of performance requirements to components (RAS).

- Specifications and the product WBS are derived from the physical architecture.

SUPPLEMENT 6-A

CONCEPT DESCRIPTION SHEET

The Concept Description Sheet describes (in textual or graphical form) the technical approach or the design concept, and shows how the system will be integrated to meet the performance and functional requirements. It is generally used in early concept design to show system concepts.

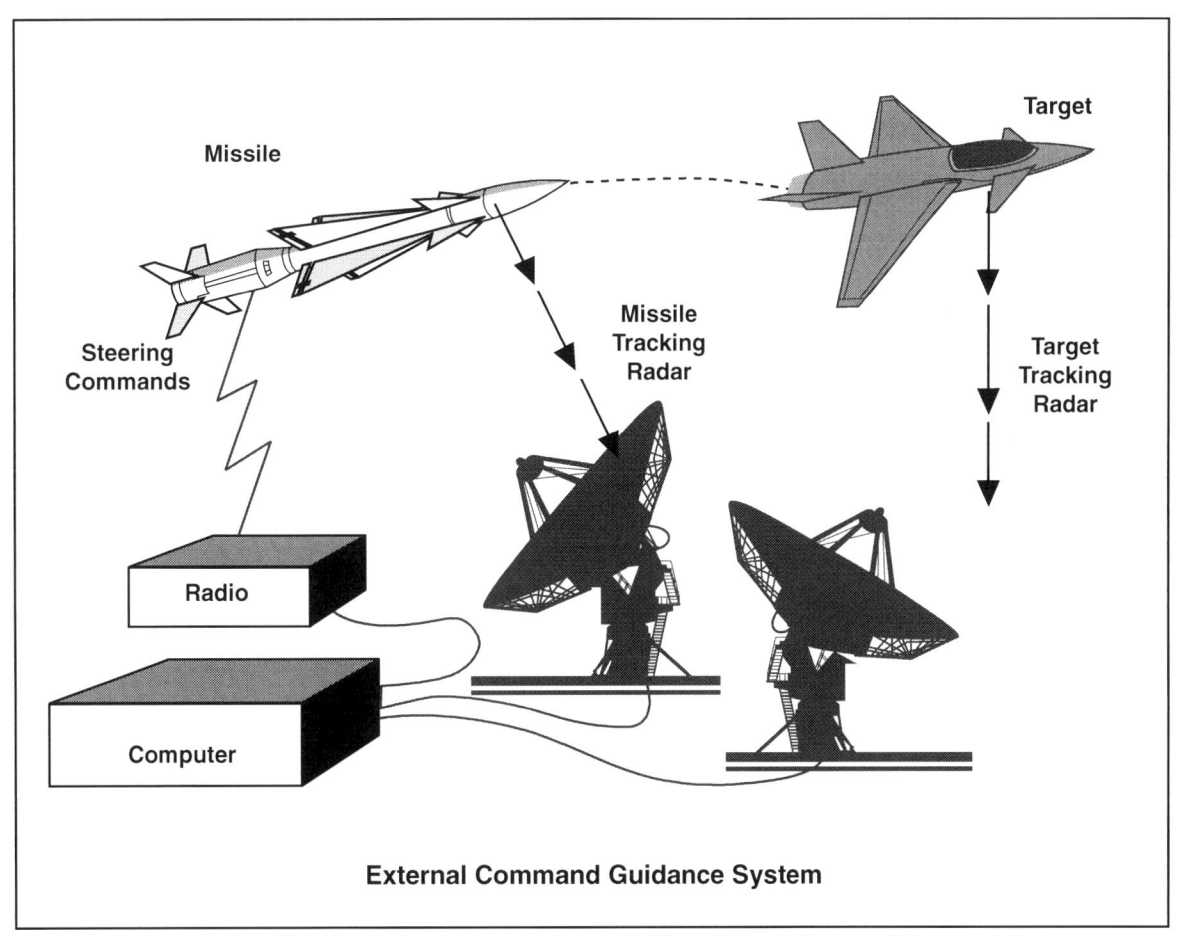

Figure 6-3. Concept Description Sheet Example

SUPPLEMENT 6-B

SCHEMATIC BLOCK DIAGRAMS

The Schematic Block Diagram (SBD) depicts hardware and software components and their interrelationships. They are developed at successively lower levels as analysis proceeds to define lower-level functions within higher-level requirements. These requirements are further subdivided and allocated using the Requirements Allocation Sheet (RAS). SBDs provide visibility of related system elements, and traceability to the RAS, FFBD, and other system engineering documentation. They describe a solution to the functional and performance requirements established by the functional architecture; show interfaces between the system components and between the system components and other systems or subsystems; support traceability between components and their functional origin; and provide a valuable tool to enhance configuration control. The SBD is also used to develop Interface Control Documents (ICDs) and provides an overall understanding of system operations.

A simplified SBD, Figure 6-4, shows how components and the connection between them are presented on the diagram. An expanded version is usually developed which displays the detailed functions performed within each component and a detailed depiction of their interrelationships. Expanded SBDs will also identify the WBS numbers associated with the components.

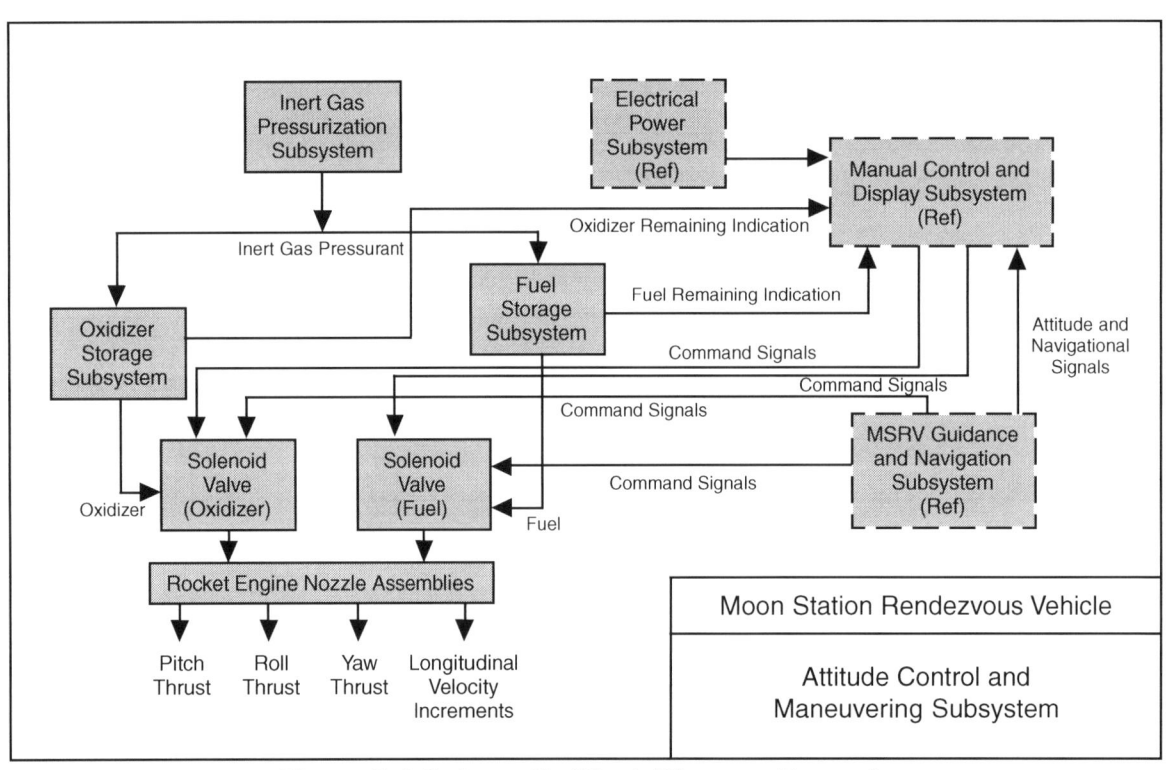

Figure 6-4. Schematic Block Diagram Example

SUPPLEMENT 6-C

REQUIREMENTS ALLOCATION SHEET

The RAS initiated in Functional Analysis and Allocation is expanded in Design Synthesis to document the connection between functional requirements and the physical system. It provides traceability between the Functional Analysis and Allocation and Synthesis activities. It is a major tool in maintaining consistency between functional architectures and the designs that are based on them. (Configuration Item (CI) numbers match the WBS.)

Requirements Allocation Sheet	Functional Flow Diagram Title and No. 2.58.4 Provide Guidance Compartment Cooling		Equipment Identification	
Function Name and No.	Functional Performance and Design Requirements	Facility Rqmnts	Nomenclature	CI or Detail Spec No.
2.58.4 Provide Guidance Compartment Cooling	The temperature in the guidance compartment must be maintained at the initial calibration temperature of +0.2 Deg F. The initial calibration temperature of the compartment will be between 66.5 and 68.5 Deg F.		Guidance Compartment Cooling System	3.54.5
2.58.4.1 Provide Chilled Coolant (Primary)	A storage capacity for 65 gal of chilled liquid coolant (deionized water) is required. The temperature of the stored coolant must be monitored continuously. The stored coolant must be maintained within a temperature range of 40–50 Deg F. for an indefinite period of time. The coolant supplied must be free of obstructive particles 0.5 micron at all times.		Guidance Compartment Coolant Storage Subsystem	3.54.5.1

Figure 6-5. Requirements Allocation Sheet (Example)

CHAPTER 7
VERIFICATION

7.1 GENERAL

The Verification process confirms that Design Synthesis has resulted in a physical architecture that satisfies the system requirements. Verification represents the intersection of systems engineering and test and evaluation.

Verification Objectives

The objectives of the Verification process include using established criteria to conduct verification of the physical architecture (including software and interfaces) from the lowest level up to the total system to ensure that cost, schedule, and performance requirements are satisfied with acceptable levels of risk. Further objectives include generating data (to confirm that system, subsystem, and lower level items meet their specification requirements) and validating technologies that will be used in system design solutions. A method to verify each requirement must be established and recorded during requirements analysis and functional allocation activities. (If it can not be verified it can not be a legitimate requirement.) The verification list should have a direct relationship to the requirements allocation sheet and be continually updated to correspond to it.

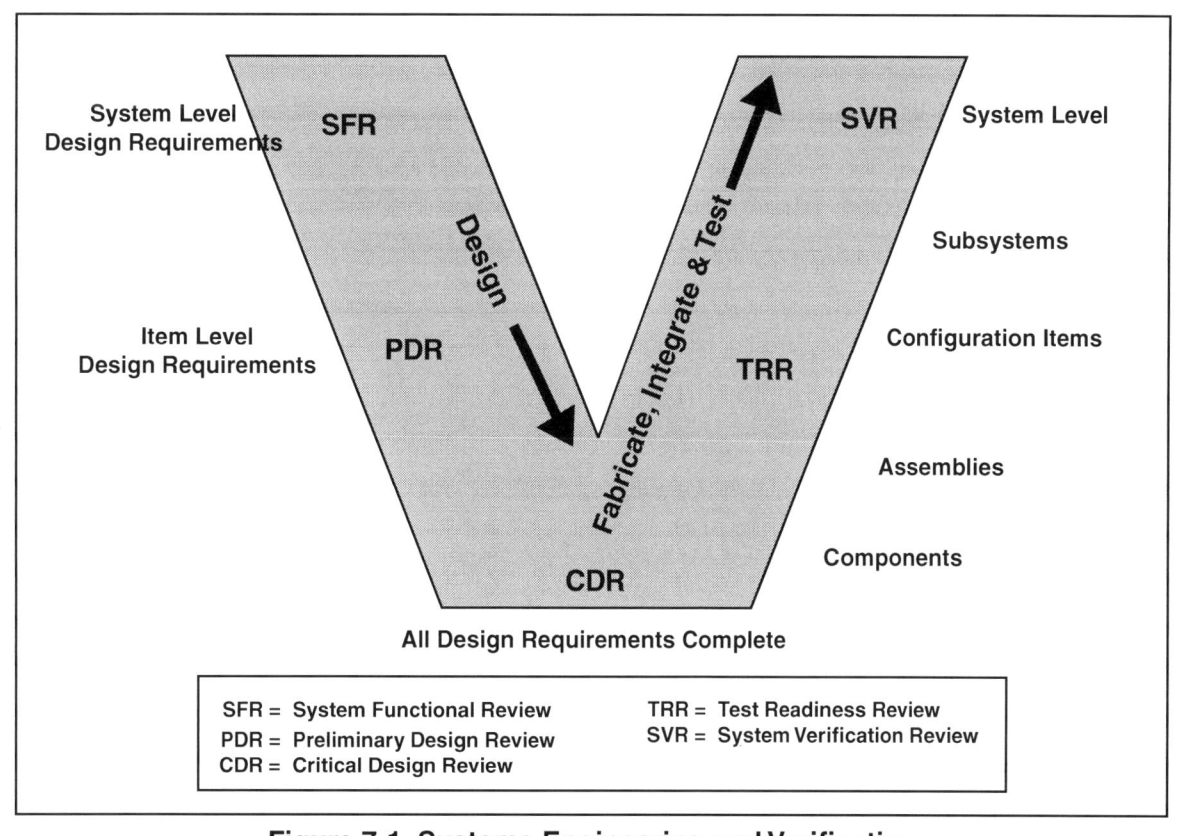

Figure 7-1. Systems Engineering and Verification

Verification Activities

System design solutions are verified by the following types of activities:

1. Analysis – the use of mathematical modeling and analytical techniques to predict the compliance of a design to its requirements based on calculated data or data derived from lower level component or subsystem testing. It is generally used when a physical prototype or product is not available or not cost effective. Analysis includes the use of both modeling and simulation which is covered in some detail in chapter 13,

2. Inspection – the visual examination of the system, component, or subsystem. It is generally used to verify physical design features or specific manufacturer identification,

3. Demonstration – the use of system, subsystem, or component operation to show that a requirement can be achieved by the system. It is generally used for a basic confirmation of performance capability and is differentiated from testing by the lack of detailed data gathering, or

4. Test – the use of system, subsystem, or component operation to obtain detailed data to verify performance or to provide sufficient information to verify performance through further analysis. Testing is the detailed quantifying method of verification, and as described later in this chapter, it is ultimately required in order to verify the system design.

Choice of verification methods must be considered an area of potential risk. Use of inappropriate methods can lead to inaccurate verification. Required defining characteristics, such as key performance parameters (KPPs) are verified by demonstration and/or test. Where total verification by test is not feasible, testing is used to verify key characteristics and assumptions used in design analysis or simulation. Validated models and simulation tools are included as analytical verification methods that complement other methods. The focus and nature of verification activities change as designs progress from concept to detailed designs to physical products.

During earlier design stages, verification focuses on proof of concept for system, subsystem and component levels. During later stages, as the product definition effort proceeds, the focus turns to verifying that the system meets the customer requirements. As shown by Figure 7-1, design is a top-down process while the Verification activity is a bottom-up process. Components will be fabricated and tested prior to the subsystems. Subsystems will be fabricated and tested prior to the completed system.

Performance Verification

Performance requirements must be objectively verifiable, i.e., the requirement must be measurable. Where appropriate, Technical Performance Measurements (TPM) and other management metrics are used to provide insight on progress toward meeting performance goals and requirements. IEEE Standard P1220 provides a structure for Verification activity. As shown in Figure 7-2 the structure is comprehensive and provides a good starting point for Verification planning.

7.2 DOD TEST AND EVALUATION

DoD Test and Evaluation (T&E) policies and procedures directly support the system engineering process of Verification. Testing is the means by which objective judgments are made regarding the extent to which the system meets, exceeds, or fails to meet stated objectives. The purpose of evaluation is to review, analyze, and assess data obtained from testing and other means to aid in making systematic decisions. The purpose of DoD T&E is to verify technical performance, operational effectiveness, operational suitability; and it provides essential information in support of decision making.

Common Types of T&E in DoD

T&E policy requires developmental tests. They confirm that technical requirements have been

Chapter 7 — Verification

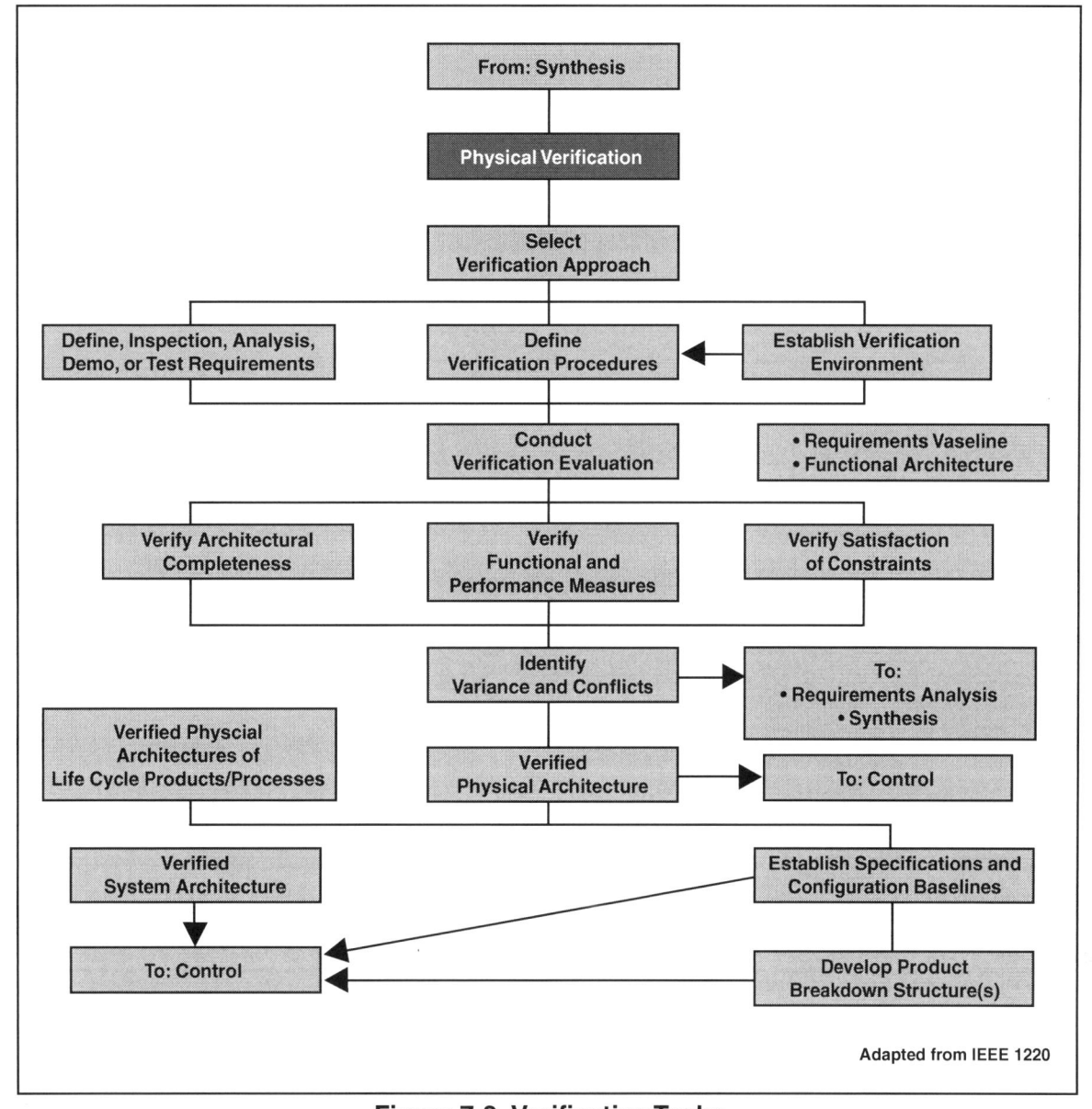

Figure 7-2. Verification Tasks (Adapted from IEEE 1220)

satisfied, and independent analysis and tests verify the system's operational effectiveness and suitability. DoD T&E traditionally and by directive is categorized as:

- Developmental T&E which focuses primarily on technical achievement,

- Operational T&E which focuses on operational effectiveness and suitability and includes Early Operational Assessments (EOA), Operational Assessment (OA), Initial Operational Test and Evaluation (IOT&E), and Follow-On Operational Test and Evaluation (FOT&E), and

- Live Fire T&E which provides assessment of the vulnerability and lethality of a system by subjecting it to real conditions comparable to the required mission.

T&E

The program office plans and manages the test effort to ensure testing is timely, efficient, comprehensive and complete—and that test results are converted into system improvements. Test planning will determine the effectiveness of the verification process. Like all systems engineering planning activities, careful attention to test planning can reduce program risk. The key test planning document is the Test and Evaluation Master Plan (TEMP). This document lays out the objectives, schedule, and resources reflecting program office and operational test organization planning decisions. To ensure integration of this effort, the program office organizes a Test Planning Work Group (TPWG) or Test Working Level IPT (WIPT) to coordinate the test planning effort.

Test Planning Work Group/Test WIPT

The TPWG/Test WIPT is intended to facilitate the integration of test requirements and activities through close coordination between the members who represent the material developer, designer community, logistic community, user, operational tester, and other stakeholders in the system development. The team outlines test needs based on system requirements, directs test design, determines needed analyses for each test, identifies potential users of test results, and provides rapid dissemination of test and evaluation results.

Test and Evaluation Master Plan (TEMP)

The Test and Evaluation Master Plan is a mandatory document prepared by the program office. The operational test organization reviews it and provides the operational test planning for inclusion. The TEMP is then negotiated between the program office and operational test organization. After differences are resolved, it is approved at appropriate high levels in the stakeholder organizations. After approval it becomes binding on managers and designers (similar to the binding nature of the Operational Requirements Document (ORD)).

The TEMP is a valuable Verification tool that provides an excellent template for technology, system, and major subsystem-level Verification planning. The TEMP includes a reaffirmation of the user requirements, and to an extent, an interpretation of what those requirements mean in various operational scenarios. Part I of the required TEMP format is *System Introduction*, which provides the mission description, threat assessment, MOEs/MOSs, a system description, and an identification of critical technical parameters. Part II, *Integrated Test Program Summary*, provides an integrated test program schedule and a description of the overall test management process. Part III, *Developmental Test & Evaluation (DT&E) Outline*, lays out an overview of DT&E efforts and a description of future DT&E. Part IV, *Operational Test & Evaluation (OT&E) Outline*, is provided by the operational test organization and includes an OT&E overview, critical operational issues, future OT&E description, and LFT&E description. Part V, *Test & Evaluation Resource Summary*, identifies the necessary physical resources and activity responsibilities. This last part includes such items as test articles, test sites, test instrumentation, test support equipment, threat representation, test targets and other expendables, operational force test support, simulations, models, test-beds, special requirements, funding, and training.

Key Performance Parameters

Every system will have a set of KPPs that are the performance characteristics that *must* be achieved by the design solution. They flow from the operational requirements and the resulting derived MOEs. They can be identified by the user, the decision authority, or the operational tester. They are documented in the TEMP.

Developmental Test and Evaluation

The DT&E verifies that the design solution meets the system technical requirements and the system is prepared for successful OT&E. DT&E activities assess progress toward resolving critical operational issues, the validity of cost-performance tradeoff decisions, the mitigation of acquisition technical risk, and the achievement of system maturity.

DT&E efforts:

- Identify potential operational and technological capabilities and limitations of the alternative concepts and design options being pursued;

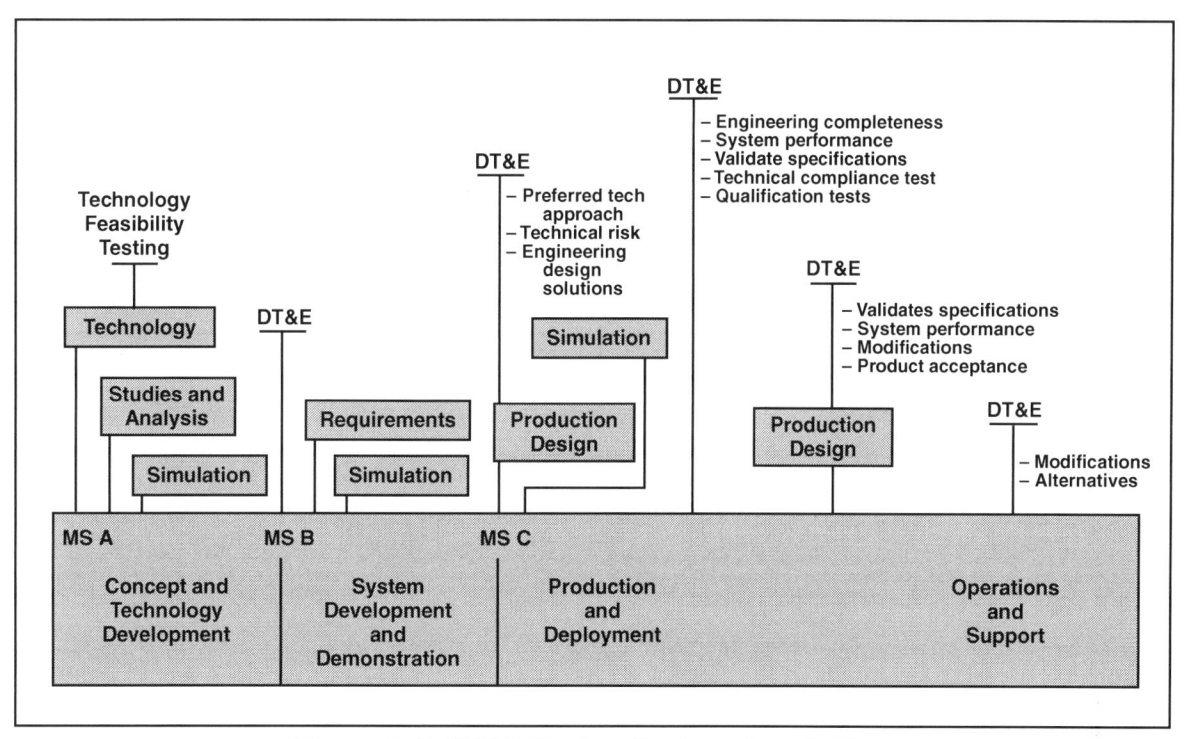

Figure 7-3. DT&E During System Acquisition

- Support the identification of cost-performance tradeoffs by providing analyses of the capabilities and limitations of alternatives;

- Support the identification and description of design technical risks;

- Assess progress toward resolving Critical Operational Issues, mitigating acquisition technical risk, achieving manufacturing process requirements and system maturity;

- Assess validity of assumptions and analysis conclusions; and

- Provide data and analysis to certify the system ready for OT&E, live-fire testing and other required certifications.

Figure 7-3 highlights some of the more significant DT&E focus areas and where they fit in the acquisition life cycle.

Live Fire Test and Evaluation

LFT&E is performed on any Acquisition Category (ACAT) I or II level weapon system that includes features designed to provide protection to the system or its users in combat. It is conducted on a production configured article to provide information concerning potential user casualties, vulnerabilities, and lethality. It provides data that can establish the system's susceptibility to attack and performance under realistic combat conditions.

Operational Test and Evaluation

OT&E programs are structured to determine the operational effectiveness and suitability of a system under realistic conditions, and to determine if the minimum acceptable operational performance requirements as specified in the ORD and reflected by the KPPs have been satisfied. OT&E uses threat-representative forces whenever possible, and employs typical users to operate and maintain the system or item under conditions simulating both combat stress and peacetime conditions. Operational tests will use production or production-

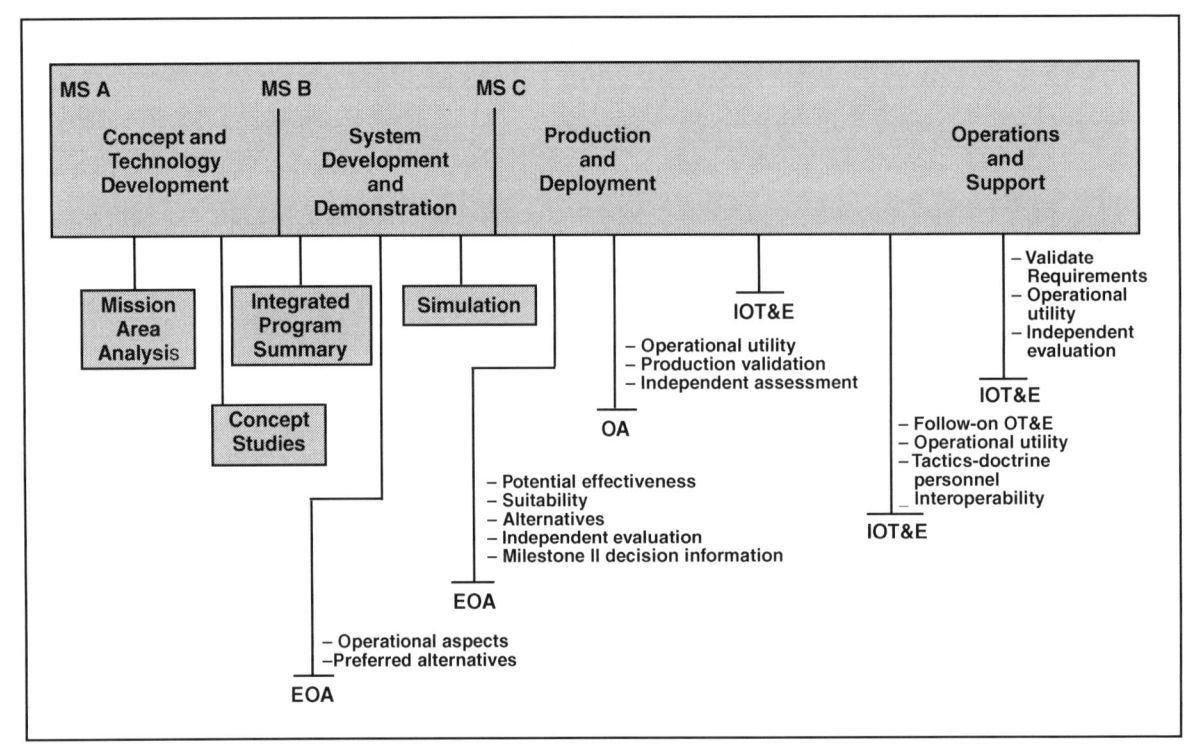

Figure 7-4. OT&E During System Acquisition

representative articles for the operational tests that support the full-rate production decision. Live Fire Tests are usually performed during the operational testing period. Figure 7-4 shows the major activities associated with operational testing and where they fit in the DoD acquisition life cycle.

OT&E Differences

Though the overall objective of both DT&E and OT&E is to verify the effectiveness and suitability of the system, there are distinct differences in their specific objects and focus. DT&E primarily focuses on verifying system technical requirements, while OT&E focuses on verifying operational requirements. DT&E is a program office responsibility that is used to develop the design. OT&E is an independent evaluation of design maturity that is used to determine if the program should proceed to full-rate production. Figure 7-5 lists the major differences between the two.

7.3 SUMMARY POINTS

The Verification activities of the Systems Engineering Process are performed to verify that physical design meets the system requirements.

- DoD T&E policy supports the verification process through a sequence of Developmental, Operational, and Live-Fire tests, analyses, and assessments. The primary management tools for planning and implementing the T&E effort are the TEMP and the integrated planning team.

Development Tests	Operational Tests
• Controlled by program manager	• Controlled by independent agency
• One-on-one tests	• Many-on-many tests
• Controlled environment	• Realistic/tactical environment with operational scenario
• Contractor environment	
• Trained, experienced operators	• No system contractor involvement
• Precise performance objectives and threshold measurements	• User troops recently trained
	• Performance measures of operational effectiveness and suitability
• Test to specification	
• Developmental, engineering, or production representative test article	• Test to operational requirements
	• Production representative test article

Figure 7-5. DT/OT Comparison

CHAPTER 8

SYSTEMS ENGINEERING PROCESS OUTPUTS

8.1 DOCUMENTING REQUIREMENTS AND DESIGNS

Outputs of the systems engineering process consist of the documents that define the system requirements and design solution. The physical architecture developed through the synthesis process is expanded to include enabling products and services to complete the system architecture. *This system level architecture then becomes the reference model for further development of system requirements and documents.* System engineering process outputs include the system and configuration item architectures, specifications, and baselines, and the decision database.

Outputs are dependent on the level of development. They become increasingly technically detailed as system definition proceeds from concept to detailed design. As each stage of system definition is achieved, the information developed forms the input for succeeding applications of the system engineering process.

Architectures: System/Configuration Item

The System Architecture describes the entire system. It includes the physical architecture produced through design synthesis and adds the enabling products and services required for life cycle employment, support, and management. Military Handbook (MIL-HDBK)-881, *Work Breakdown Structures*, provides reference models for weapon systems architectures. As shown by Figure 8-1, MIL-HDBK-881 illustrates the first three levels of typical system architectures. Program Offices can use MIL-HDBK-881 templates during system definition to help develop a top-level architecture tailored to the needs of the specific system considered. The design contractor will normally develop the levels below these first three. Chapter 9 of this text describes the WBS in more detail.

Specifications

A specification is a document that clearly and accurately describes the essential technical requirements for items, materials, or services including the procedures by which it can be determined that the requirements have been met. Specifications help avoid duplication and inconsistencies, allow for accurate estimates of necessary work and resources, act as a negotiation and reference document for engineering changes, provide documentation of configuration, and allow for consistent communication among those responsible for the eight primary functions of Systems Engineering. They provide IPTs a precise idea of the problem to be solved so that they can efficiently design the system and estimate the cost of design alternatives. They provide guidance to testers for verification (qualification) of each technical requirement.

Program-Unique Specifications

During system development a series of specifications are generated to describe the system at different levels of detail. These program unique specifications form the core of the configuration baselines. As shown by Figure 8-2, in addition to referring to different levels within the system hierarchy, these baselines are defined at different phases of the design process.

Initially the system is described in terms of the top-level (system) functions, performance, and interfaces. These technical requirements are derived from the operational requirements established by

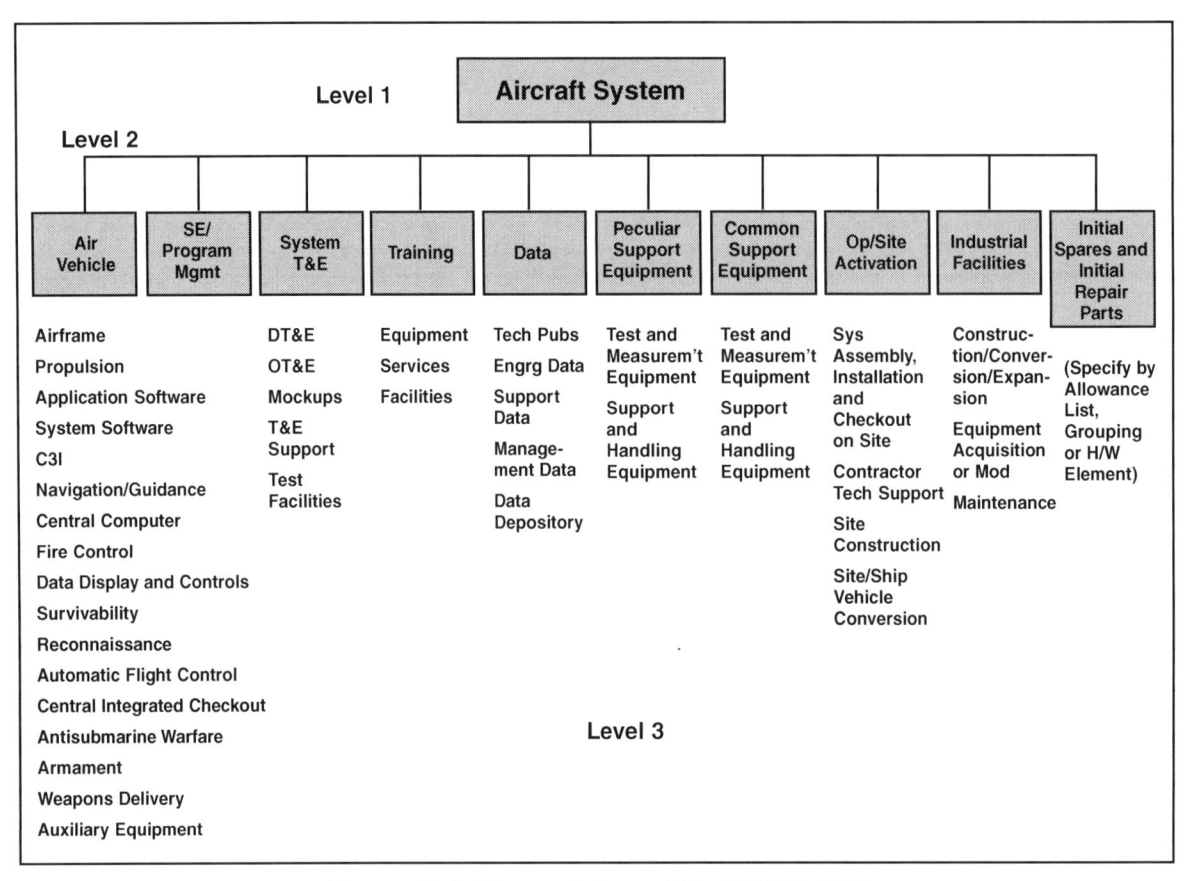

Figure 8-1. Example from MIL-HDBK-881

the user. This system-level technical description is documented in the System Specification, which is the primary documentation of the system-level Functional Baseline. The system requirements are then flowed down (allocated) to the items below the system level, such that a set of design criteria are established for each of those items. These item descriptions are captured in a set of Item Performance Specifications, which together with other interface definitions, process descriptions, and drawings, document the Allocated Baseline (sometimes referred to as the "Design To" baseline). Having baselined the design requirements for the individual items, detailed design follows. Detailed design involves defining the system from top to bottom in terms of the physical entities that will be employed to satisfy the design requirements. When detailed design is complete, a final baseline is defined. This is generally referred to as the Product Baseline, and, depending on the stage of development, may reflect a "Build to" or "As built" description. The Product Baseline is documented by the Technical Data Package, which will include not only Item Detail Specifications, but also, Process and Material Specifications, as well as drawings, parts lists, and other information that describes the final system in full physical detail. Figure 8-3 shows how these specifications relate to their associated baselines.

Role of Specifications

Requirements documents express why the development is needed. Specification documents are an intermediate expression of what the needed system has to do in terms of technical requirements (function, performance, and interface). Design documents (drawings, associated lists, etc.) describe the means by which the design requirements are to be satisfied. Figure 8-4 illustrates how requirements flow down from top-level specifications to design documentation. Preparation of specifications are part of the system engineering process, but also involve techniques that relate to

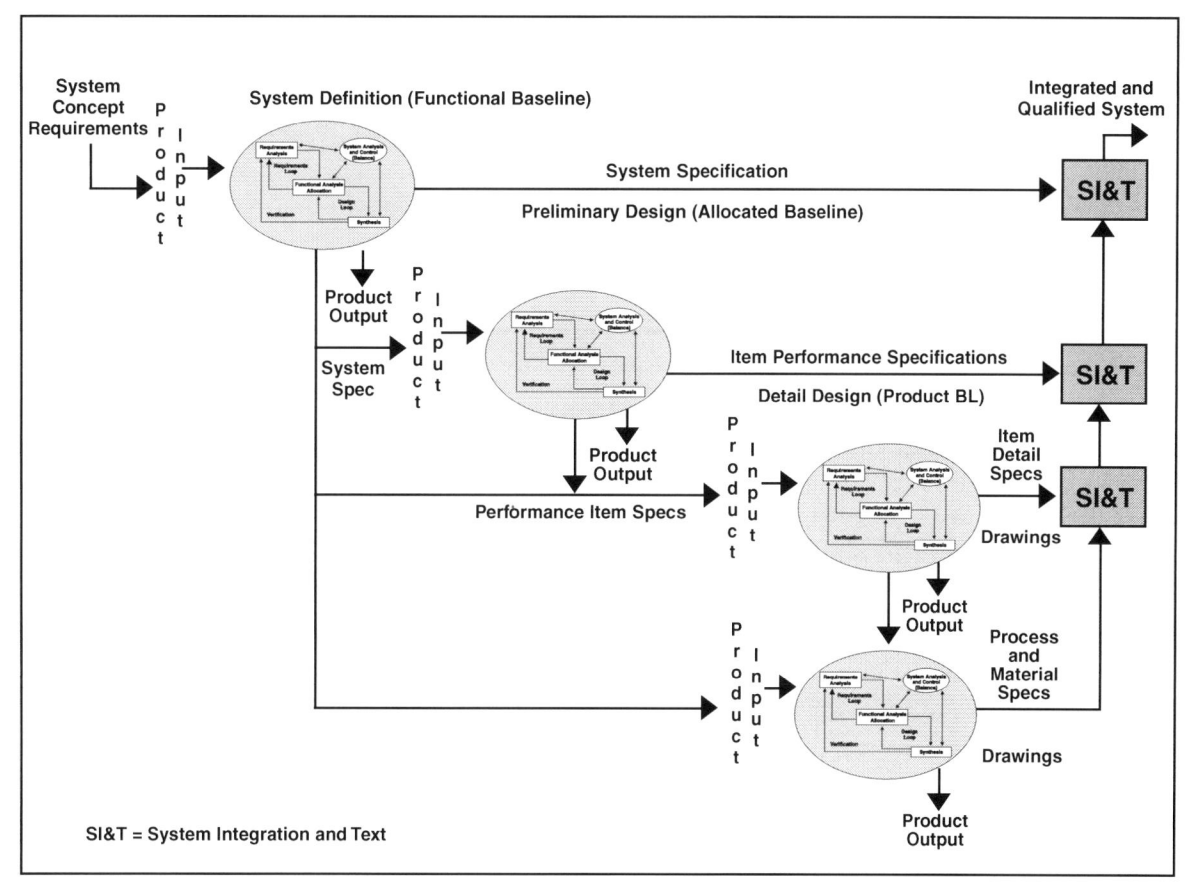

Figure 8-2. Specifications and Levels of Development

Specification	Content	Baseline
System Spec	Defines mission/technical performance requirements. Allocates requirements to functional areas and defines interfaces.	Functional
Item Performance Spec	Defines performance characteristics of CIs and CSCIs. Details design requirements and with drawings and other documents form the Allocated Baseline.	Allocated "Design To"
Item Detail Spec	Defines form, fit, function, performance, and test requirements for acceptance. (Item, process, and material specs start the Product Baseline effort, but the final audited baseline includes all the items in the TDP.)	Product "Build To" or "As Built"
Process Spec	Defines process performed during fabrication.	
Material Spec	Defines production of raw materials or semi-fabricated material used in fabrication.	

Figure 8-3. Specification Types

communication skills, both legal and editorial. Figure 8-5 provides some rules of thumb that illustrate this.

In summary, specifications document what the system has to do, how well it has to do it, and how to verify it can do it.

Baselines

Baselines formally document a product at some given level of design definition. They are references for the subsequent development to follow. Most DoD systems are developed using the three classic baselines described above: functional, allocated, and product. Though the program unique specifications are the dominant baseline documentation, they alone do not constitute a baseline.

Additional documents include both end and enabling product descriptions. End product baseline documents normally include those describing system requirements, functional architecture, physical architecture, technical drawing package, and requirements traceability. Enabling product baseline documents include a wide range of documents that could include manufacturing plans and processes, supportability planning, supply documentation, manuals, training plans and programs, test planning, deployment planning, and others. All enabling products should be reviewed for their susceptibility to impact from system configuration changes. If a document is one that describes a part of a system and could require change if the configuration changes, then most likely it should be included as a baseline document.

Acquisition Program Baselines

Acquisition Program Baselines and Configuration Baselines are related. To be accurate the Program baseline must reflect the realities of the Configuration Baseline, but the two should not be confused. Acquisition Program Baselines are high level assessments of program maturity and viability. Configuration Baselines are system descriptions. Figure 8-6 provides additional clarification.

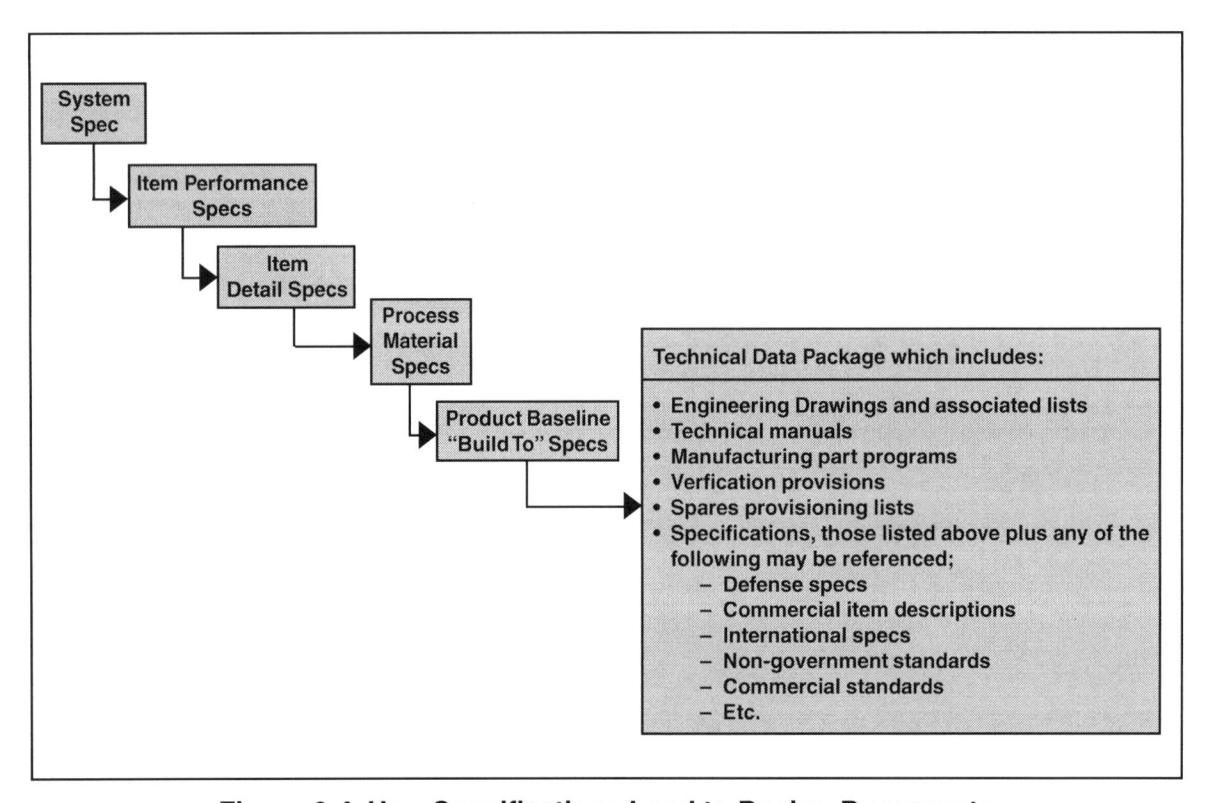

Figure 8-4. How Specifications Lead to Design Documents

- Use a table of contents and define all abbreviations and acronyms.
- Use active voice.
- Use "shall" to denote mandatory requirement and "may" or "should" to denote guidance provisions.
- Avoid ambiguous provisions, such as "as necessary," "contractor's best practice," "smooth finish," and similar terms.
- Use the System Engineering Process to identify requirements. Do not over-specify.
- Avoid "tiering." Any mandatory requirement in a document below the first tier, should be stated in the specification.
- Only requirement sections of the MIL-STD-491D formats are binding. Do not put requirements in non-binding sections, such as *Scope*, *Documents*, or *Notes*.
- Data documentation requirements are specified in a Contract Data Requirements List.

Figure 8–5. Rules of Thumb for Specification Preparation

Decision Database

The decision database is the documentation that supports and explains the configuration solution decisions. It includes trade studies, cost effectiveness analyses, Quality Function Deployment (QFD) analysis, models, simulations, and other data generated to understand a requirement, develop alternative solutions, or make a choice between them. These items are retained and controlled as part of the Data Management process described in Chapter 10.

8.2 DOD POLICY AND PRACTICE—SPECIFICATIONS AND STANDARDS

DoD uses specifications to communicate product requirements and standards to provide guidance concerning proven methods and practices.

Specifications

DoD uses three basic classifications of specifications: materiel specifications (developed by DoD components), Program-Unique Specifications, and non-DoD specifications.

- Program Baselines
    - Embody only the most important cost, schedule, and performance objectives and thresholds
    - Threshold breach results in re-evaluation of program at MDA level
    - Selected key performance parameters
    - Specifically evolves over the development cycle and is updated at each major milestone review or program restructure
- Required on ALL programs for measuring and reporting status

- Configuration Baselines
  Identify and define an item's functional and physical characteristics
    - *Functional Baseline* – Describes system level requirements
    - *Allocated Baseline* – Describes design requirements for items below system level
    - *Product Baseline* – Describes product physical detail
- Documents outputs of Systems Engineering Process

Figure 8–6. Acquisition Program Baselines and Configuration Baselines

DoD developed specifications describe essential technical requirements for purchase of materiel. Program-Unique Specifications are an integral part of the system development process. Standard practice for preparation of DoD and Program-Unique Specifications is guided by MIL-STD-961D. This standard provides guidance for the development of performance and detail specifications. MIL-STD-961D, Appendix A provides further guidance for the development of Program-Unique Specifications.

Non-DoD specifications and standards approved for DoD use are listed in the *DoD Index of Specifications and Standards* (DoDISS).

DoD Policy (Specifications)

DoD policy is to develop *performance* specifications for procurement and acquisition. In general, detail specifications are left for contractor development and use. Use of a detail specification in DoD procurement or acquisition should be considered only where absolutely necessary, and then only with supporting trade studies and acquisition authority approval.

DoD policy gives preference to the use of commercial solutions to government requirements, rather than development of unique designs. Therefore, the use of commercial item specifications and descriptions should be a priority in system architecture development. Only when no commercial solution is available should government detail specifications be employed.

In the case of re-procurement, where detail specifications and drawings are government owned, standardization or interface requirements may present a need for use of detailed specifications. Trade studies that reflect total ownership costs and the concerns related to all eight primary functions should govern decisions concerning the type of specification used for re-procurement of systems, subsystems, and configuration items. Such trade studies and cost analysis should be preformed prior to the use of detail specifications or the decision to develop and use performance specifications in a reprocurement.

Performance Specifications

Performance Specifications state requirements in terms of the required results with criteria for verifying compliance, but without stating the methods for achieving the required results. In general, performance specifications define products in terms of functions, performance, and interface requirements. They define the functional requirements for the item, the environment in which it must operate, and interface and interchangeability characteristics. The contractor is provided the flexibility to decide how the requirements are best achieved, subject to the constraints imposed by the government, typically through interface requirements. System Specifications and Item Performance Specifications are examples of performance specifications.

Detail Specifications

Detail Specifications, such as Item Detail, Material and Process Specifications, provide design requirements. This can include materials to be used, how a requirement is to be achieved, or how an item is to be fabricated or constructed. If a specification contains both performance and detail requirements, it is considered a Detail Specification, with the following exception: Interface and interchangeability requirements in Performance Specifications may be expressed in detailed terms. For example, a Performance Specification for shoes would specify size requirements in detailed terms, but material or method of construction would be stated in performance terms.

Software Documentation – IEEE/EIA 12207

IEEE/EIA 12207, *Software Life Cycle Processes*, describes the U.S. implementation of the ISO standard on software processes. This standard describes the development of software specifications as one aspect of the software development process.

The process described in IEEE/EIA 12207 for allocating requirements in a top-down fashion and documenting the requirements at all levels parallels the systems engineering process described in this text. The standard requires first that system-level requirements be allocated to software items (or configuration items) and that the software requirements then be documented in terms of functionality, performance, and interfaces, and that qualification requirements be specified. Software item requirements must be traceable to system-level, and be consistent and verifiable.

The developer is then required to decompose each software item into software components and then into software units that can be coded. Requirements are allocated from item level, to component, and finally to unit level. This is the detailed design activity and IEEE/EIA 12207 requires that these allocations of requirements be documented in documents that are referred to as "descriptions," or, if the item is a "stand alone" item, as "specifications." The content of these documents is defined in the IEEE/EIA standard; however, the level of detail required will vary by project. Each project must therefore ensure that a common level of expectation is established among all stakeholders in the software development activity.

Standard Practice for Defense Specifications – MIL-STD-961D

The purpose of MIL-STD-961D is to establish uniform practices for specification preparation, to ensure inclusion of essential requirements, to ensure Verification (qualification) methods are established for each requirement, and to aid in the use and analysis of specification content. MIL-STD-961D establishes the format and content of system, configuration item, software, process and material specifications. These Program-Unique Specifications are developed through application of the systems engineering process and represent a hierarchy as shown in Figure 8-7.

Standards

Standards establish engineering and technical limitations and applications for items, materials, processes, methods, designs, and engineering practices. They are "corporate knowledge" documents describing how to do some process or a

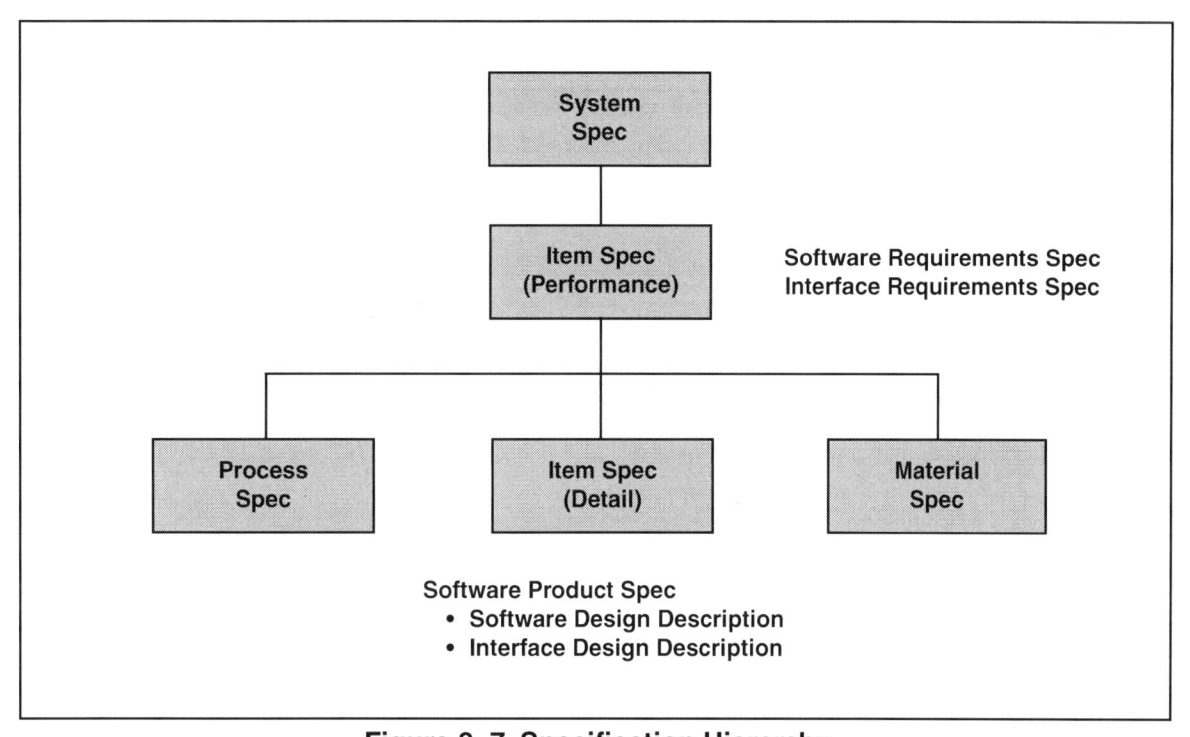

Figure 8–7. Specification Hierarchy

description of a body of knowledge. Standards come from many sources, reflecting the practices or knowledge base of the source. Format and content of Defense Standards, including Handbooks, are governed by MIL-STD-962. Other types of standards in use in DoD include Commercial Standards, Corporate Standards, International Standards, Federal Standards, and Federal Information Processing Standards.

DoD Policy (Standards)

DoD policy does not require standard management approaches or manufacturing processes on contracts. This policy applies to the imposition of both Military Specifications and Standards and, in addition, to the imposition of Commercial and Industry Standards. In general, the preferred approach is to allow contractors to use industry, government, corporate, or company standards they have determined to be appropriate to meet government's needs. The government reviews and accepts the contractor's approach through a contract selection process or a contractual review process.

The government should impose a process or standard only as a last resort, and only with the support of an appropriate trade study analysis. If a specific standard is imposed in a solicitation or contract, a waiver will be required from an appropriate Service authority.

However, there is need on occasion to direct the use of some standards for reasons of standardization, interfaces, and development of open systems. A case in point is the mandated use of the Joint Technical Architecture (JTA) for defining interoperability standards. The JTA sets forth the set of interface standards that are expected to be employed in DoD systems. The JTA is justifiably mandatory because it promotes needed interoperability standardization, establishes supportable interface standards, and promotes the development of open systems.

DoD technical managers should be alert to situations when directed standards are appropriate to their program. Decisions concerning use of directed standards should be confirmed by trade studies and requirements traceability.

DoD Index of Specifications and Standards

The DoDISS lists all international, adopted industry standardization documents authorized for use by the military departments, federal and military specifications and standards. Published in three volumes, it contains over 30,000 documents in 103 Federal Supply Groups broken down into 850 Federal Supply Classes. It covers the total DoD use of specifications and standards, ranging from fuel specifications to international quality standards.

8.3 SUMMARY POINTS

- System Engineering Process Outputs include the system/configuration item architecture, specifications and baselines, and the decision database.

- System/Configuration Item Architectures include the physical architecture and the associated products and services.

- Program-Unique specifications are a primary output of the System Engineering Process. Program-Unique specifications describe what the system or configuration item must accomplish and how it will be verified. Program-Unique specifications include the System, Item Performance, and Item Detail Specifications. The System Specification describes the system requirements, while Item Performance and Item Detail Specifications describe configuration item requirements.

- Configuration baselines are used to manage and control the technical development. Program baselines are used for measuring and supporting program status.

- The Decision Database includes those documents or software that support understanding and decision making during formulation of the configuration baselines.

- DoD policy is to develop *performance* specifications for procurement and acquisition. Use of other than performance specifications in a contract must be justified and approved.

- It is DoD policy not to require standard management approaches or manufacturing processes on contracts.

- Mandatory use of some standard practices are necessary, but must be justified through analysis. A case in point is the mandatory use of the standards listed in the Joint Technical Architecture.

PART 3

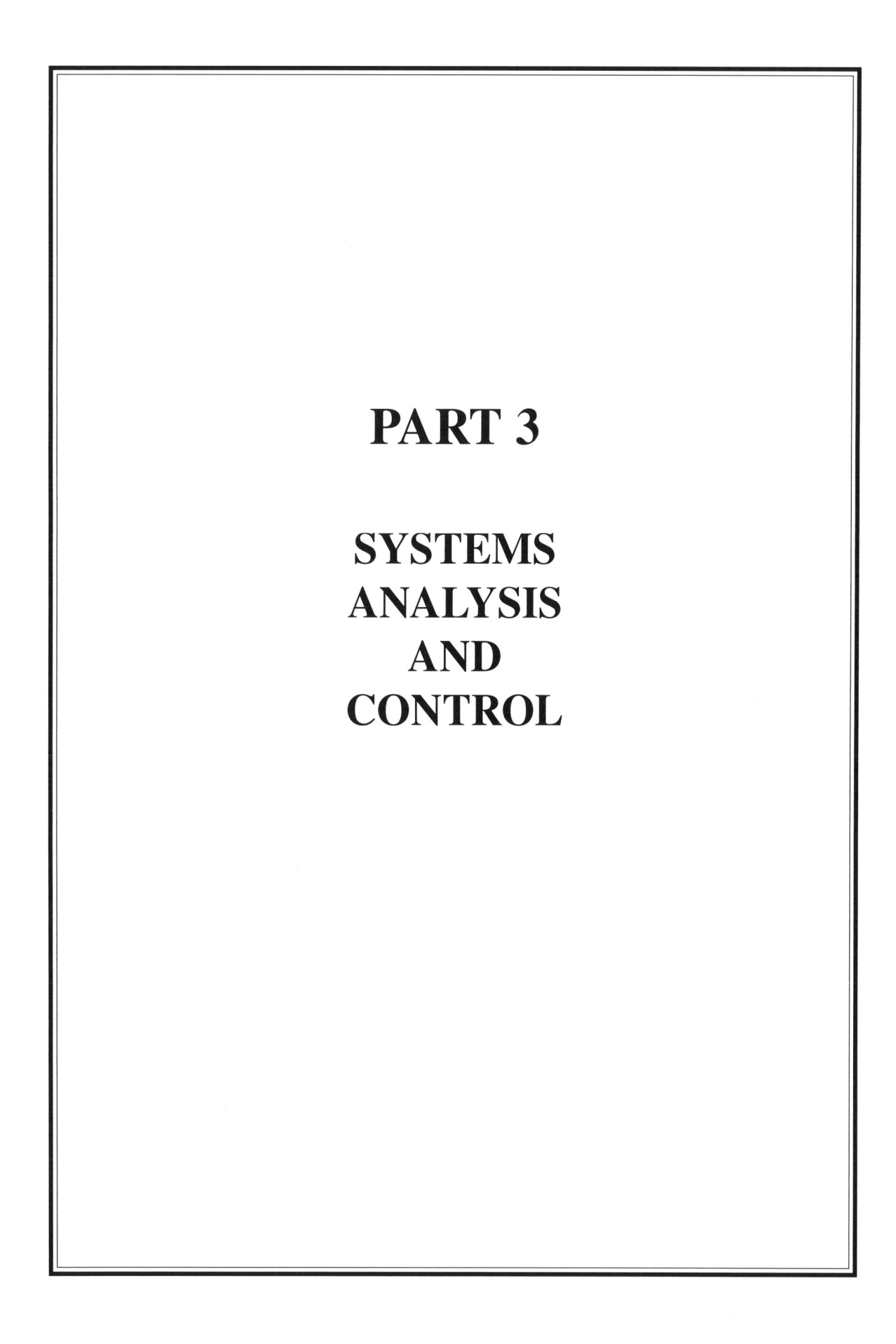

SYSTEMS ANALYSIS AND CONTROL

CHAPTER 9

WORK BREAKDOWN STRUCTURE

9.1 INTRODUCTION

The Work Breakdown Structure (WBS) is a means of organizing system development activities based on system and product decompositions. The systems engineering process described in earlier chapters produces system and product descriptions. These product architectures, together with associated services (e.g., program management, systems engineering, etc.) are organized and depicted in a hierarchical tree-like structure that is the WBS. (See Figure 9-1.)

Because the WBS is a direct derivative of the physical and systems architectures it could be considered an output of the systems engineering process. It is being presented here as a Systems Analysis and Control tool because of its essential utility for all aspects of the systems engineering process. It is used to structure development activities, to identify data and documents, and to organize integrated teams, and for other non-technical program management purposes.

WBS Role in DoD Systems Engineering

DoD 5000.2-R requires that a program WBS be established to provide a framework for program and technical planning, cost estimating, resource allocation, performance measurement, and status reporting. The WBS is used to define the total system, to display it as a product-oriented family tree composed of hardware, software, services, data, and facilities, and to relate these elements to each other and to the end product. Program offices are to tailor a program WBS using the guidance provided in MIL-HDBK-881.

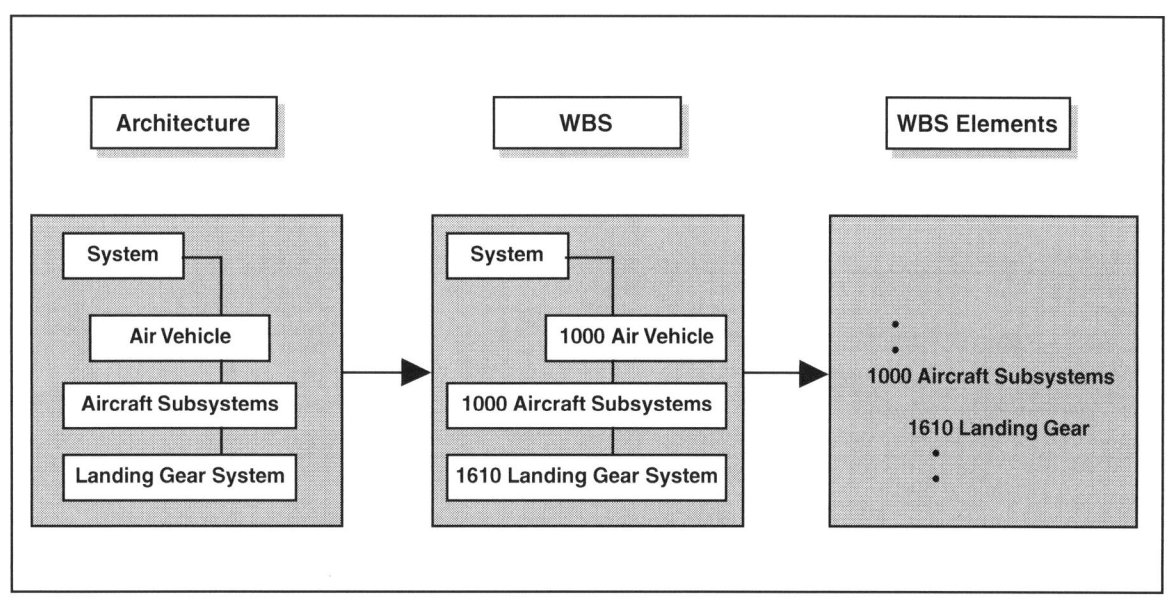

Figure 9-1. Architecture to WBS Flow

The program WBS is developed initially to define the top three levels. As the program proceeds through development and is further defined, program managers should ensure that the WBS is extended to identify all high-cost and high-risk elements for management and reporting, while ensuring the contractor has complete flexibility to extend the WBS below the reporting requirement to reflect how work will be accomplished.

Basic Purposes of the WBS

Organizational:
The WBS provides a coordinated, complete, and comprehensive view of program management. It establishes a structure for organizing system development activities, including IPT design, development, and maintenance.

Business:
It provides a structure for budgets and cost estimates. It is used to organize collection and analysis of detailed costs for earned value reports (Cost Performance Reports or Cost/Schedule Control System Criteria reporting).

Technical:
The WBS establishes a structure for:

- Identifying products, processes, and data,

- Organizing risk management analysis and tracking,

- Enabling configuration and data management. It helps establish interface identification and control.

- Developing work packages for work orders and material/part ordering, and

- Organizing technical reviews and audits.

The WBS is used to group product items for specification development, to develop Statements of Work (SOW), and to identify specific contract deliverables.

WBS – Benefits

The WBS allows the total system to be described through a logical breakout of product elements into work packages. A WBS, correctly prepared, will account for all program activity. It links program objectives and activities with resources, facilitates initial budgets, and simplifies subsequent cost reporting. The WBS allows comparison of various independent metrics and other data to look for comprehensive trends.

It is a foundation for all program activities, including program and technical planning, event schedule definition, configuration management, risk management, data management, specification preparation, SOW preparation, status reporting and problem analysis, cost estimates, and budget formulation.

9.2 WBS DEVELOPMENT

The physical and system architectures are used to prepare the WBS. The architectures should be reviewed to ensure that all necessary products and services are identified, and that the top-down structure provides a continuity of flow down for all tasks. Enough levels must be provided to identify work packages for cost/schedule control purposes. If too few levels are identified, then management visibility and integration of work packages may suffer. If too many levels are identified, then program review and control actions may become excessively time-consuming.

The first three WBS Levels are organized as:
    Level 1 – Overall System
    Level 2 – Major Element (Segment)
    Level 3 – Subordinate Components (Prime
         Items)

Levels below the first three represent component decomposition down to the configuration item level. In general, the government is responsible for the development of the first three levels, and the contractor(s) for levels below three.

DoD Practice

In accordance with DoD mandatory procedures in DoD 5000.2-R and common DoD practice as established in MIL-HDBK-881, the program office develops a program WBS and a contract WBS for each contract. The program WBS is the WBS that represents the total system, i.e., the WBS that describes the system architecture. The contract WBS is the part of the program WBS that relates to deliverables and tasks of a specific contract.

MIL-HDBK-881 is used by the program office to support the systems engineering process in developing the first three levels of the program WBS, and to provide contractors with guidance for lower level WBS development. As with most standards and handbooks, use of MIL-HDBK-881 cannot be specified as a contract requirement.

Though WBS development is a systems engineering activity, it impacts cost and budget professionals, as well as contracting officers. An integrated team representing these stakeholders should be formed to support WBS development.

WBS Anatomy

A program WBS has an end product part and an enabling product part. The end product part of the system typically consists of the prime mission product(s) delivered to the operational customer. This part of the WBS is based on the physical architectures developed from operational requirements. It represents that part of the WBS involved in product development. Figure 9-2 presents a simple example of a program WBS product part.

The "enabling product" part of the system includes the products and services required to develop, produce, and support the end product(s). This part of the WBS includes the horizontal elements of the system architecture (exclusive of the end products), and identifies all the products and services necessary to support the life cycle needs of the product. Figure 9-3 shows an example of the top three levels of a complete WBS tree.

Contract WBS

A contract WBS is developed by the program office in preparation for contracting for work required to develop the system. It is further developed by the contractor after contract award. The contract WBS is that portion of the program WBS that is specifically being tasked through the contract. A simple example of a contract WBS derived from the program WBS shown in Figure 9-2 is provided by Figure 9-4. Figure 9-4, like Figure 9-2, only includes the product part of the contract WBS. A

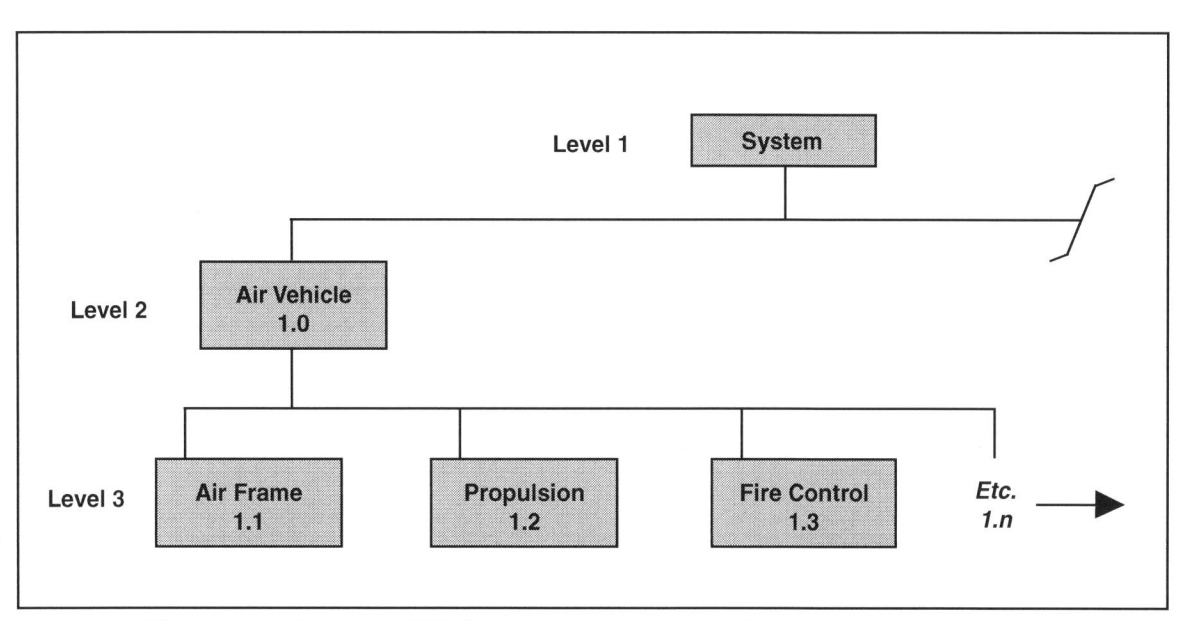

Figure 9-2. Program WBS – The Product Part (Physical Architecture)

Systems Engineering Fundamentals   Chapter 9

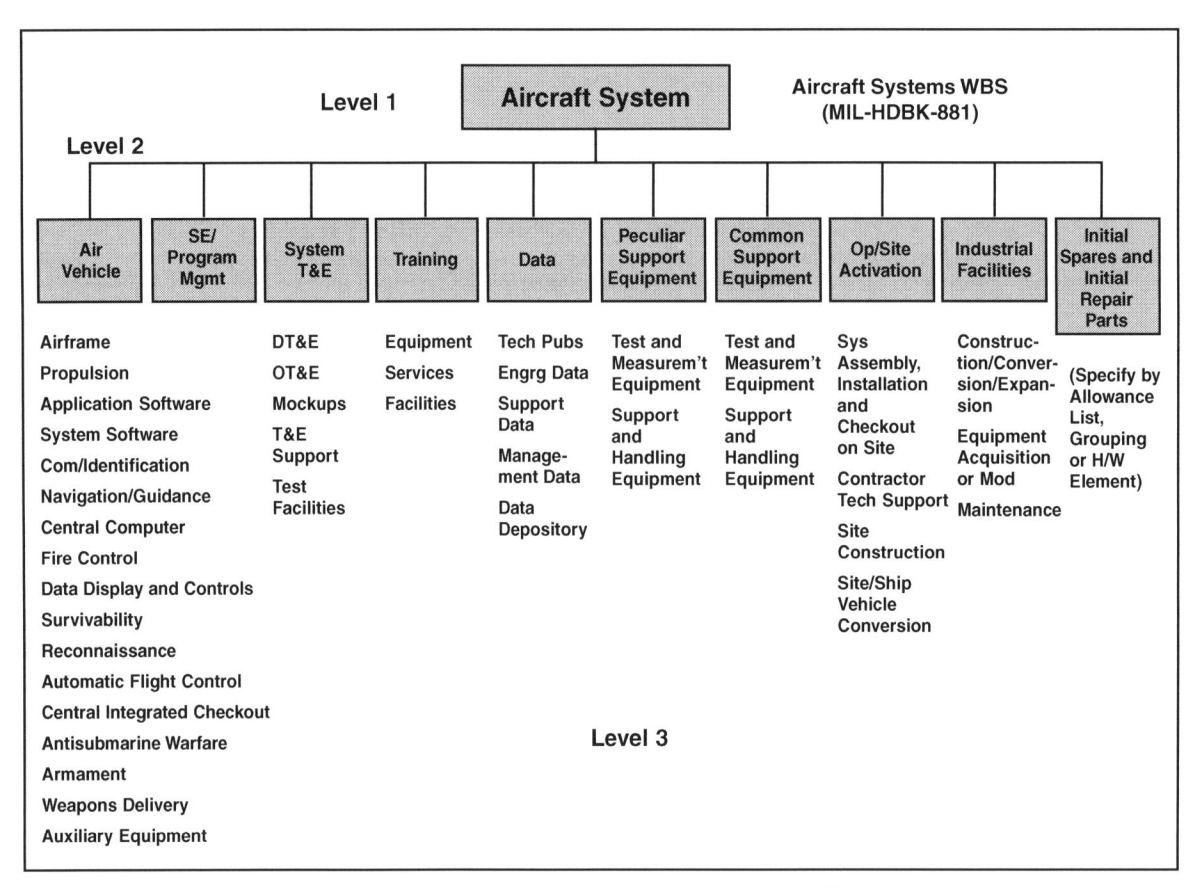

```
                                                    Aircraft Systems WBS
                        Level 1    [Aircraft System]    (MIL-HDBK-881)
       Level 2
    ┌──────┬──────┬──────┬──────┬──────┬──────┬──────┬──────┬──────┐
   Air    SE/    System Training Data  Peculiar Common Op/Site Industrial Initial
  Vehicle Program T&E                  Support  Support Activation Facilities Spares and
          Mgmt                         Equipment Equipment                    Initial
                                                                              Repair
                                                                              Parts

  Airframe          DT&E      Equipment Tech Pubs  Test and    Test and    Sys         Construc-     (Specify by
  Propulsion        OT&E      Services  Engrg Data Measurem't  Measurem't  Assembly,   tion/Conver-  Allowance
  Application Software Mockups Facilities Support  Equipment   Equipment   Installation sion/Expan-  List,
  System Software   T&E                  Data      Support     Support     and         sion          Grouping
  Com/Identification Support             Manage-   and         and         Checkout    Equipment     or H/W
  Navigation/Guidance Test               ment Data Handling    Handling    on Site     Acquisition   Element)
  Central Computer  Facilities           Data      Equipment   Equipment   Contractor  or Mod
  Fire Control                           Depository                        Tech Support Maintenance
  Data Display and Controls                                                Site
  Survivability                                                            Construction
  Reconnaissance                                                           Site/Ship
  Automatic Flight Control                                                 Vehicle
  Central Integrated Checkout                           Level 3            Conversion
  Antisubmarine Warfare
  Armament
  Weapons Delivery
  Auxiliary Equipment
```

Figure 9-3. The Complete Work Breakdown Structure

complete contract WBS would include associated enabling products, similar to those identified in Figure 9-3. The resulting complete contract WBS is used to organize and identify contractor tasks. The program office's preliminary version is used to develop a SOW for the Request for Proposals.

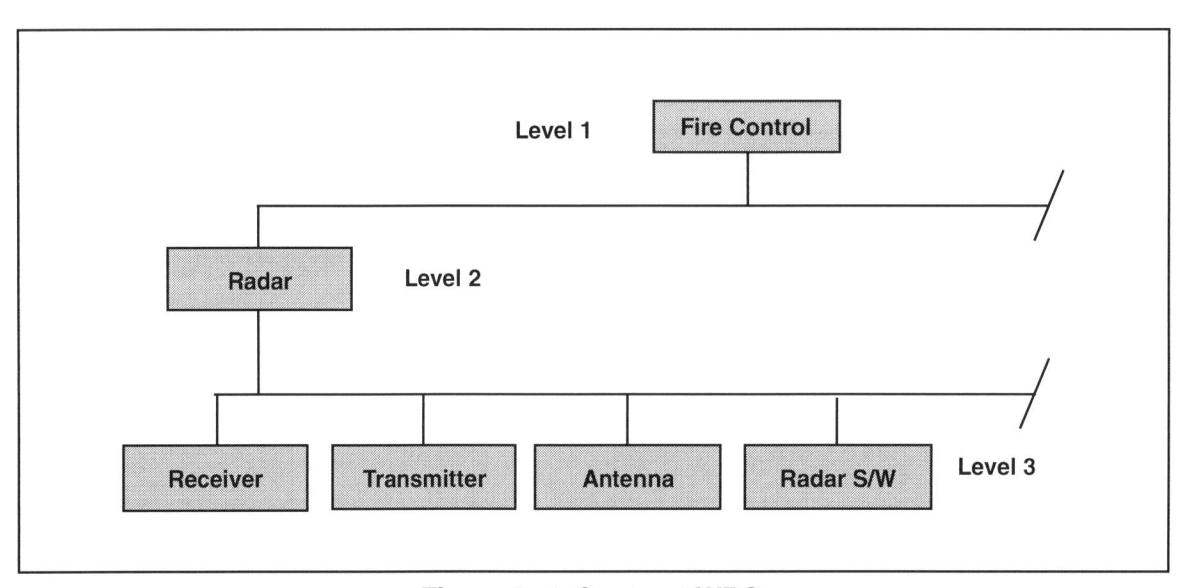

```
                    Level 1    [Fire Control]
                                     │
                                     ├─────────────/
                                     │
                    [Radar]   Level 2
                       │
                       ├─────────────/
        ┌──────────┬──────────┬──────────┬──────────┐
    [Receiver] [Transmitter] [Antenna] [Radar S/W]   Level 3
```

Figure 9–4. Contract WBS

88

9.3 DESIGNING AND TRACKING WORK

A prime use of the WBS is the design and tracking of work. The WBS is used to establish what work is necessary, a logical decomposition down to work packages, and a method for organizing feedback. As shown by Figure 9-5, the WBS element is matrixed against those organizations in the company responsible for the task. This creates cost accounts and task definition at a detailed level. It allows rational organization of integrated teams and other organizational structures by helping establish what expertise and functional support is required for a specific WBS element. It further allows precise tracking of technical and other management.

WBS Dictionary

As part of the work and cost control use of the WBS, a Work Breakdown Dictionary is developed. For each WBS element a dictionary entry is prepared that describes the task, what costs (activities) apply, and the references to the associated Contract Line Item Numbers and SOW paragraph. An example of a level 2 WBS element dictionary entry is shown as Figure 9-6.

9.4 SUMMARY POINTS

- The WBS is an essential tool for the organization and coordination of systems engineering

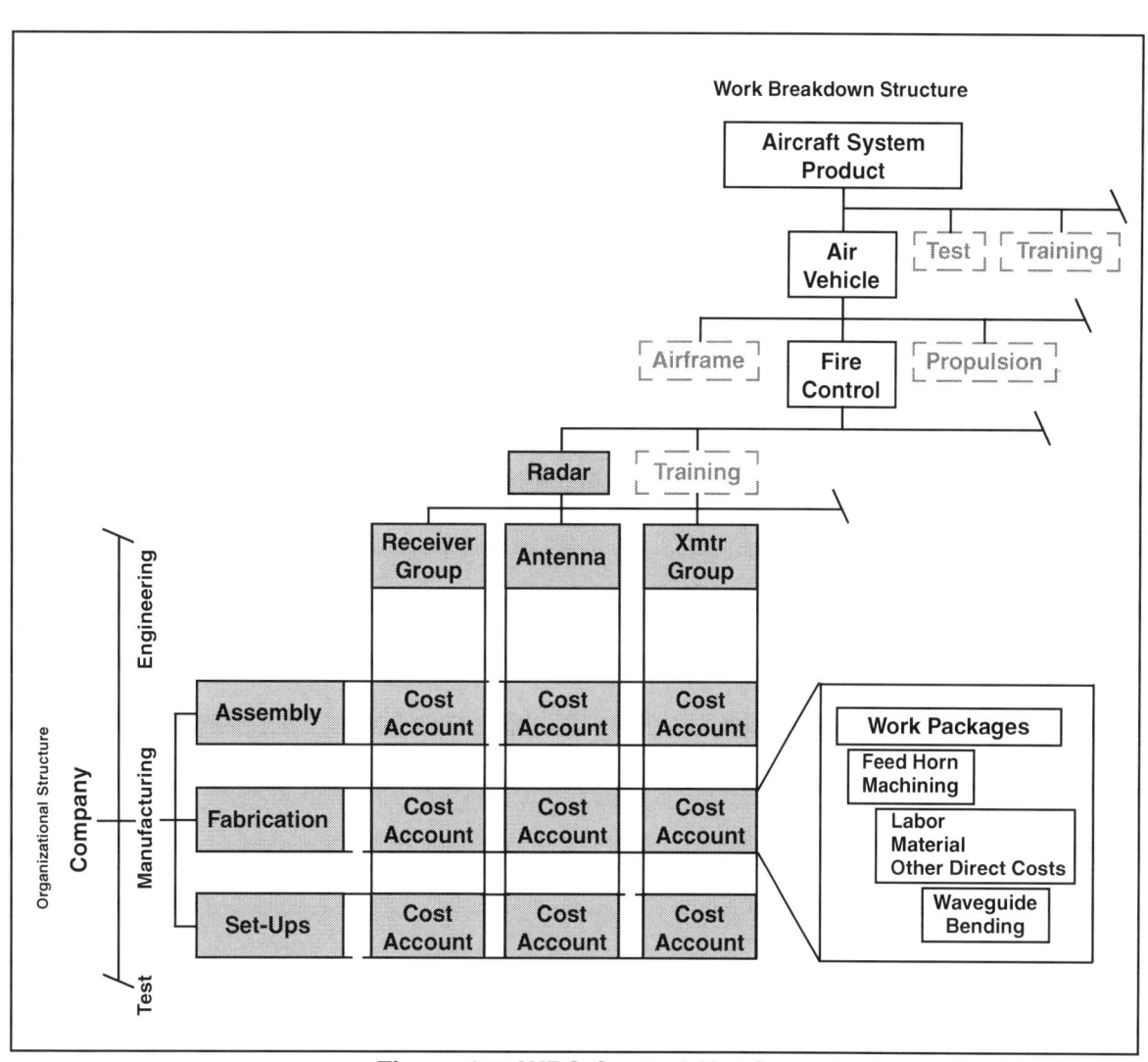

Figure 9-5. WBS Control Matrix

Index Item No. 2		WBS Level 2		CONTRACT NUMBER F33657-72-C-0923
WBS Element A10100		**WBS Title** Air Vehicle		Contract Line Item: 0001, 0001AA, 0001AB, 0001AC, 0001AD 0001AE, 0001AF, 0001AG, 0001AH
Date Chg	Revision No.	Revision Auth	Approved	
Specification No. 689E078780028	\multicolumn{3}{l|}{**Specification Title:** Prime Item Development Specification for AGM 86A Air Vehicle/ Airframe}	

Element Task Description	Cost Description

Technical Content:
The Air Vehicle element task description refers to the effort required to develop, fabricate, integrate and test the airframe segment, portions of the Navigation/Guidance element, and Airborne Development Test Equipment and Airborne Operational Test Equipment and to the integration assembly and check-out of these complete elements, together with the Engine Segment, to produce the complete Air Vehicle. The lower-level elements included and summarized in the Air Vehicle element are:
    Airframe Segment (A11100), Navigation/Guidance Segment (A32100), Airborne Development Test Equipment (A61100), and Airborne Operational Test Equipment (A61200).
	MPC/PMC Work Order/Work Auth
	A10100 See lower level
	                                         WBS Elements
	
	Cost Content – System Contractor
	The cost to be accumulated against this element includes a summarization of all costs required to plan, develop, fabricate, assemble, integrate and perform development testing, analysis and reporting for the air vehicle. It also includes all costs associated with the required efforts in integrating, assembling and checking our GFP required to create this element.
	                            Applicable SOW Paragraph
	                                            3.6.2

Figure 9-6. Work Breakdown Dictionary

processes, and it is a product of the systems engineering process.

- Its importance extends beyond the technical community to business professionals and contracting officers. The needs of all stakeholders must be considered in its development. The program office develops the program WBS and a high-level contract WBS for each contract. The contractors develop the lower levels of the contract WBS associated with their contract.

- The system architecture provides the structure for a program WBS. SOW tasks flow from this WBS.

- The WBS provides a structure for organizing IPTs and tracking metrics.

CHAPTER 10

CONFIGURATION MANAGEMENT

10.1 FOUNDATIONS

Configuration Defined

A "configuration" consists of the functional, physical, and interface characteristics of existing or planned hardware, firmware, software or a combination thereof as set forth in technical documentation and ultimately achieved in a product. The configuration is formally expressed in relation to a Functional, Allocated, or Product configuration baseline as described in Chapter 8.

Configuration Management

Configuration management permits the orderly development of a system, subsystem, or configuration item. A good configuration management program ensures that designs are traceable to requirements, that change is controlled and documented, that interfaces are defined and understood, and that there is consistency between the product and its supporting documentation. Configuration management provides documentation that describes what is supposed to be produced, what is being produced, what has been produced, and what modifications have been made to what was produced.

Configuration management is performed on baselines, and the approval level for configuration modification can change with each baseline. In a typical system development, customers or user representatives control the operational requirements and usually the system concept. The developing agency program office normally controls the functional baseline. Allocated and product baselines can be controlled by the program office, the producer, or a logistics agent depending on the life cycle management strategy. This sets up a hierarchy of configuration control authority corresponding to the baseline structure. Since lower level baselines have to conform to a higher-level baseline, changes at the lower levels must be examined to assure they do not impact a higher-level baseline. If they do, they must be approved at the highest level impacted. For example, suppose the only engine turbine assembly affordably available for an engine development cannot provide the continuous operating temperature required by the allocated baseline. Then not only must the impact of the change at the lower level (turbine) be examined, but the change should also be reviewed for possible impact on the functional baseline, where requirements such as engine power and thrust might reside.

Configuration management is supported and performed by integrated teams in an Integrated Product and Process Development (IPPD) environment. Configuration management is closely associated with technical data management and interface management. Data and interface management is essential for proper configuration management, and the configuration management effort has to include them.

DoD Application of Configuration Management

During the development contract, the Government should maintain configuration control of the functional and performance requirements only, giving contractors responsibility for the detailed design. (SECDEF Memo of 29 Jun 94.) This implies government control of the Functional (system requirements) Baseline. Decisions regarding whether or not the government will take control of the lower-level baselines (allocated and product baselines), and when ultimately depends on the requirements and strategies needed for the particular program. In general, government control of lower-level baselines, if exercised, will take place late in the development program after design has stabilized.

Configuration Management Planning

When planning a configuration management effort you should consider the basics: what has to be done, how should it be done, who should do it, when should it be done, and what resources are required. Planning should include the organizational and functional structure that will define the methods and procedures to manage functional and physical characteristics, interfaces, and documents of the system component. It should also include statements of responsibility and authority, methods of control, methods of audit or verification, milestones, and schedules. EIA IS-649, National Consensus Standard for Configuration Management, and MIL-HDBK-61 can be used as planning guidance.

Configuration Item (CI)

A key concept that affects planning is the configuration item (CI). CI decisions will determine what configurations will be managed. CIs are an aggregation of hardware, firmware, or computer software, or any of their discrete portions, which satisfies an end-use function and is designated for separate configuration management. Any item required for logistic support and designated for separate procurement is generally identified as CI. Components can be designated CIs because of crucial interfaces or the need to be integrated with operation with other components within or outside of the system. An item can be designated CI if it is developed wholly or partially with government funds, including nondevelopmental items (NDI) if additional development of technical data is required. All CIs are directly traceable to the WBS.

Impact of CI Designation

CI designation requires a separate configuration management effort for the CI, or groupings of related CIs. The decision to place an item, or items, under formal configuration control results in:

- Separate specifications,

- Formal approval of changes,

- Discrete records for configuration status accounting,

- Individual design reviews and configuration audits,

- Discrete identifiers and name plates,

- Separate qualification testing, and

- Separate operating and user manuals.

10.2 CONFIGURATION MANAGEMENT STRUCTURE

Configuration management comprises four interrelated efforts:

- Identification,

- Control,

- Status Accounting, and

- Audits.

Also directly associated with configuration management are data management and interface management. Any configuration management planning effort must consider all six elements.

Identification

Configuration Identification consists of documentation of formally approved baselines and specifications, including:

- Selection of the CIs,

- Determination of the types of configuration documentation required for each CI,

- Documenting the functional and physical characteristics of each CI,

- Establishing interface management procedures, organization, and documentation,

- Issuance of numbers and other identifiers associated with the system/CI configuration structure, including internal and external interfaces, and

- Distribution of CI identification and related configuration documentation.

Configuration Documentation

Configuration documentation is technical documentation that identifies and defines the item's functional and physical characteristics. It is developed, approved, and maintained through three distinct evolutionary increasing levels of detail. The three levels of configuration documentation form the three baselines and are referred to as functional, allocated, and product configuration documentation. These provide the specific technical description of a system or its components at any point in time.

Configuration Control

Configuration Control is the systematic proposal, justification, prioritization, evaluation, coordination, approval or disapproval, and implementation of all approved changes in the configuration of a system/CI after formal establishment of its baseline. In other words, it is how a system (and its CIs) change control process is executed and managed.

Configuration Control provides management visibility, ensures all factors associated with a proposed change are evaluated, prevents unnecessary or marginal changes, and establishes change priorities. In DoD it consists primarily of a change process that formalizes documentation and provides a management structure for change approval.

Change Documents Used for Government Controlled Baselines

There are three types of change documents used to control baselines associated with government configuration management: Engineering Change Proposal, Request for Deviation, and Request for Waivers.

- Engineering Change Proposals (ECP) identify need for a permanent configuration change. Upon approval of an ECP a new configuration is established.

- Requests for Deviation or Waiver propose a temporary departure from the baseline. They allow for acceptance of non-conforming material. After acceptance of a deviation or waiver the documented configuration remains unchanged.

Engineering Change Proposal (ECP)

An ECP is documentation that describes and suggests a change to a configuration baseline. Separate ECPs are submitted for each change that has a distinct objective. To provide advanced notice and reduce paperwork, Preliminary ECPs or Advance Change/Study Notices can be used preparatory to issue of a formal ECP. Time and effort for the approval process can be further reduced through use of joint government and contractor integrated teams to review and edit preliminary change proposals.

ECPs are identified as Class I or Class II. Class I changes require government approval before changing the configuration. These changes can result from problems with the baseline requirement, safety, interfaces, operating/servicing capability, preset adjustments, human interface including skill level, or training. Class I changes can also be used to upgrade already delivered systems to the new configuration through use of retrofit, mod kits, and the like. Class I ECPs are also used to change contractual provisions that do not directly impact the configuration baseline; for example, changes affecting cost, warranties, deliveries, or

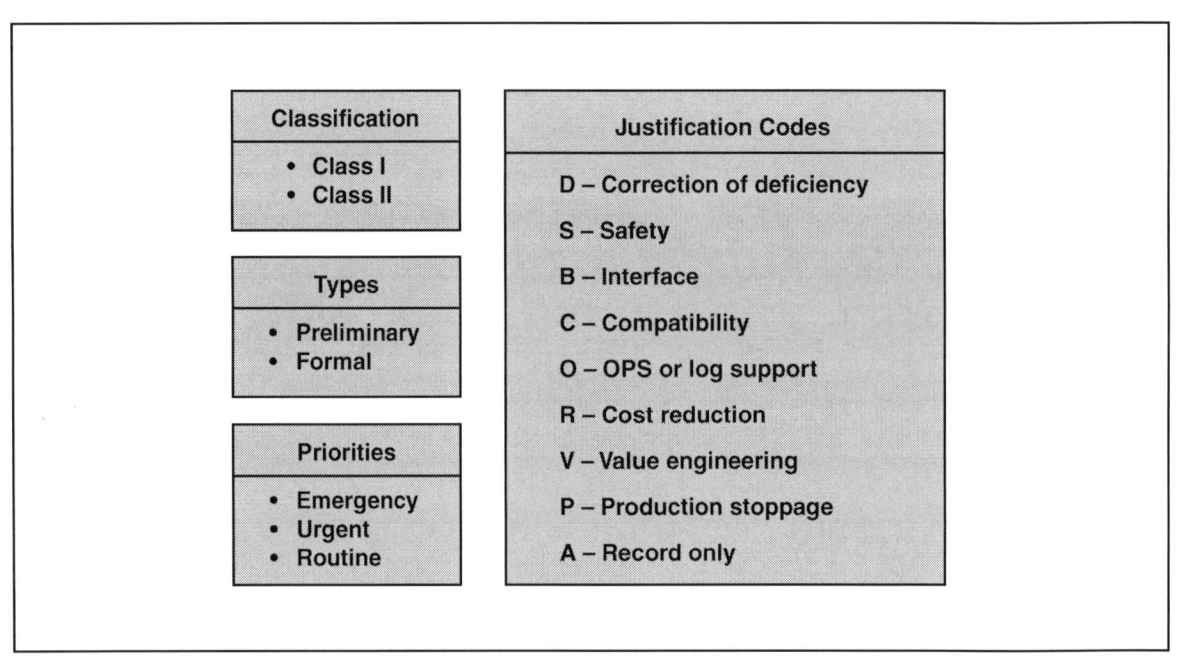

```
┌─────────────────────────────────────────────────────────────────┐
│                                                                 │
│              ┌──────────────────┐   ┌──────────────────────┐    │
│              │  Classification  │   │  Justification Codes │    │
│              ├──────────────────┤   ├──────────────────────┤    │
│              │  • Class I       │   │ D – Correction of    │    │
│              │  • Class II      │   │      deficiency      │    │
│              └──────────────────┘   │ S – Safety           │    │
│                                     │ B – Interface        │    │
│              ┌──────────────────┐   │ C – Compatibility    │    │
│              │      Types       │   │ O – OPS or log       │    │
│              ├──────────────────┤   │      support         │    │
│              │  • Preliminary   │   │ R – Cost reduction   │    │
│              │  • Formal        │   │ V – Value            │    │
│              └──────────────────┘   │      engineering     │    │
│                                     │ P – Production       │    │
│              ┌──────────────────┐   │      stoppage        │    │
│              │    Priorities    │   │ A – Record only      │    │
│              ├──────────────────┤   └──────────────────────┘    │
│              │  • Emergency     │                                │
│              │  • Urgent        │                                │
│              │  • Routine       │                                │
│              └──────────────────┘                                │
│                                                                 │
└─────────────────────────────────────────────────────────────────┘
```

Figure 10-1. ECP Designators

data requirements. Class I ECPs require program office approval, which is usually handled through a formal Configuration Control Board, chaired by the government program manager or delegated representative.

Class II changes correct minor conflicts, typos, and other "housekeeping" changes that basically correct the documentation to reflect the current configuration. Class II applies only if the configuration is not changed when the documentation is changed. Class II ECPs are usually handled by the in-plant government representative. Class II ECPs generally require only that the government concurs that the change is properly classified. Under an initiative by the Defense Contract Management Command (DCMC), contractors are increasingly delegated the authority to make ECP classification decisions.

Figure 10-1 shows the key attributes associated with ECPs. The preliminary ECP, mentioned in Figure 10-1, is a simplified version of a formal ECP that explains the proposed ECP, and establishes an approximate schedule and cost for the change. The expense of an ECP development is avoided if review of the Preliminary ECP indicates the change is not viable. The approach used for preliminary ECPs vary in their form and name. Both Preliminary ECPs and Advanced Change/Study Notices have been used to formalize this process, but forms tailored to specific programs have also been used.

Configuration Control Board (CCB)

A CCB is formed to review Class I ECPs for approval, and make a recommendation to approve or not approve the proposed change. The CCB chair, usually the program manager, makes the final decision. Members advise and recommend, but the authority for the decision rests with the chair. CCB membership should represent the eight primary functions with the addition of representation of the procurement office, program control (budget), and Configuration Control manager, who serves as the CCB secretariat.

The CCB process is shown in Figure 10-2. The process starts with the contractor. A request to the contractor for an ECP or Preliminary ECP is necessary to initiate a government identified configuration change. The secretariat's review process includes assuring appropriate government

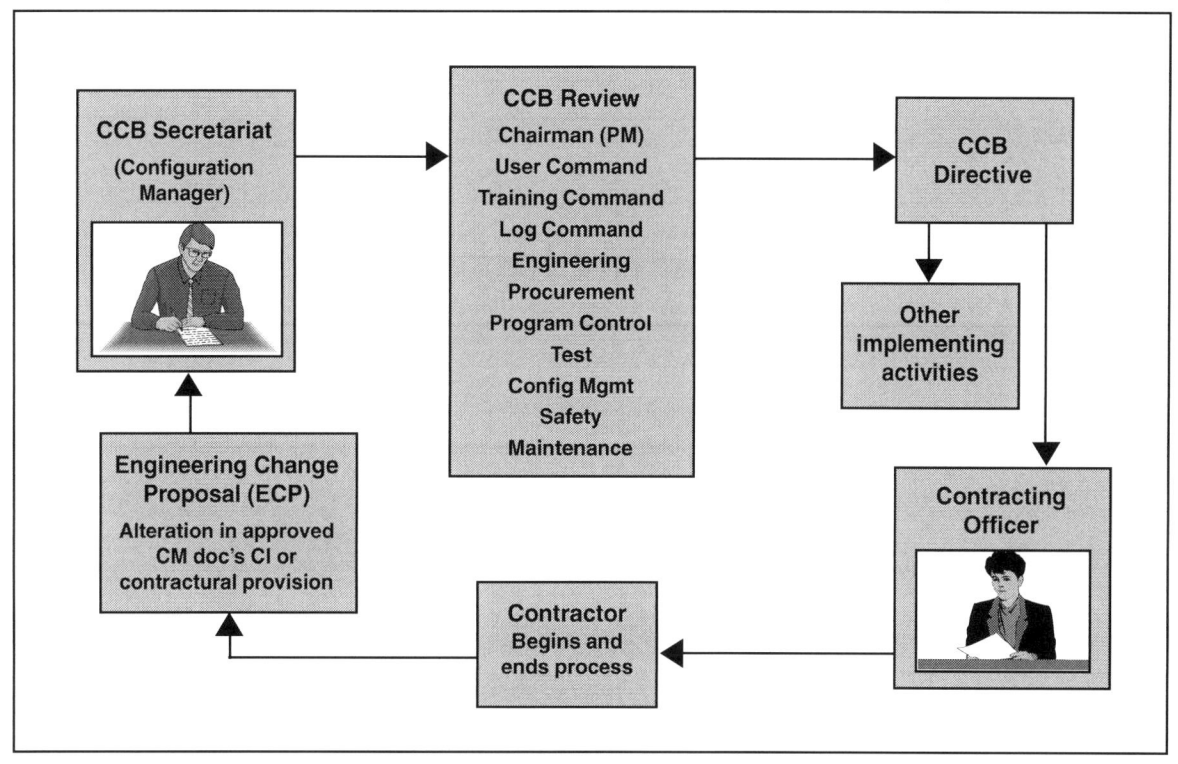

Figure 10-2. Configuration Control Board

contractual and engineering review is done prior to receipt by the CCB.

CCB Management Philosophy

The CCB process is a configuration control process, but it is also a contractual control process. Decisions made by the CCB chair affects the contractual agreement and program baseline as well as the configuration baseline. Concerns over contractual policy, program schedule, and budget can easily come into conflict with concerns relating to configuration management, technical issues, and technical activity scheduling. The CCB technical membership and CCB secretariat is responsible to provide a clear view of the technical need and the impact of alternate solutions to these conflicts. The CCB secretariat is further responsible to see that the CCB is fully informed and prepared, including ensuring that:

- A government/contractor engineering working group has analyzed the ECP and supporting data, prepared comments for CCB consideration, and is available to support the CCB;

- All pertinent information is available for review;

- The ECP has been reviewed by appropriate functional activities; and

- Issues have been identified and addressed.

CCB Documentation

Once the CCB chair makes a decision concerning an ECP, the CCB issues a Configuration Control Board Directive that distributes the decision and identifies key information relating to the implementation of the change:

- Implementation plan (who does what when);

- Contracts affected (prime and secondary);

- Dates of incorporation into contracts;

- Documentation affected (drawings, specifications, technical manuals, etc.), associated cost, and schedule completion date; and

- Identification of any orders or directives needed to be drafted and issued.

Request for Deviation or Waiver

A deviation is a specific written authorization, granted prior to manufacture of an item, to depart from a performance or design requirement for a specific number of units or a specific period of time.

A waiver is a written authorization to accept a CI that departs from specified requirements, but is suitable for use "as is" or after repair.

Requests for deviation and waivers relate to a temporary baseline departure that can affect system design and/or performance. The baseline remains unchanged and the government makes a determination whether the alternative "non-conforming" configuration results in an acceptable substitute. Acceptable substitute usually implies that there will be no impact on support elements, systems affected can operate effectively, and no follow-up or correction is required. The Federal Acquisition Regulations (FAR) requires "consideration" on government contracts when the Government accepts a "non-conforming" unit.

The distinction between Request for Deviation and Request for a Waiver is that a deviation is used *before* final assembly of the affected unit, and a waiver is used *after* final assembly or acceptance testing of the affected unit.

Status Accounting

Configuration Status Accounting is the recording and reporting of the information that is needed to manage the configuration effectively, including:

- A listing of the approved configuration documentation,

- The status of proposed changes, waivers and deviations to the configuration identification,

- The implementation status of approved changes, and

- The configuration of all units, including those in the operational inventory.

Purpose of Configuration Status Accounting

Configuration Status Accounting provides information required for configuration management by:

- Collecting and recording data concerning:
  – Baseline configurations,
  – Proposed changes, and
  – Approved changes.

- Disseminating information concerning:
  – Approved configurations,
  – Status and impact of proposed changes,
  – Requirements, schedules, impact and status of approved changes, and
  – Current configurations of delivered items.

Audits

Configuration Audits are used to verify a system and its components' conformance to their configuration documentation. Audits are key milestones in the development of the system and do not stand alone. The next chapter will show how they fit in the overall process of assessing design maturity.

Functional Configuration Audits (FCA) and the System Verification Review (SVR) are performed in the Production Readiness and LRIP stage of the Production and Development Phase. FCA is used to verify that actual performance of the configuration item meets specification requirements. The SVR serves as system-level audit after FCAs have been conducted.

The Physical Configuration Audit (PCA) is normally held during Rate Production and Development stage as a formal examination of a production representative unit against the draft technical data package (product baseline documentation).

Most audits, whether FCA or PCA, are today approached as a series of "rolling" reviews in which items are progressively audited as they are produced such that the final FCA or PCA becomes

significantly less oppressive and disruptive to the normal flow of program development.

10.3 INTERFACE MANAGEMENT

Interface Management consists of identifying the interfaces, establishing working groups to manage the interfaces, and the group's development of interface control documentation. Interface Management identifies, develops, and maintains the external and internal interfaces necessary for system operation. It supports the configuration management effort by ensuring that configuration decisions are made with full understanding of their impact outside of the area of the change.

Interface Identification

An interface is a functional, physical, electrical, electronic, mechanical, hydraulic, pneumatic, optical, software, or similar characteristic required to exist at a common boundary between two or more systems, products, or components. Normally, in a contractual relationship the procuring agency identifies external interfaces, sets requirements for integrated teams, and provides appropriate personnel for the teams. The contracted design agent or manufacturer manages internal interfaces; plans, organizes, and leads design integrated teams; maintains internal and external interface requirements; and controls interfaces to ensure accountability and timely dissemination of changes.

Interface Control Working Group (ICWG)

The ICWG is the traditional forum to establish official communications link between those responsible for the design of interfacing systems or components. Within the IPPD framework ICWGs can be integrated teams that establish linkage between interfacing design IPTs, or could be integrated into a system-level engineering working group. Membership of ICWGs or comparable integrated teams should include membership from each contractor, significant vendors, and participating government agencies. The procuring program office (external and selected top-level interfaces) or prime contractor (internal interfaces) generally designates the chair.

Interface Control Documentation (ICD)

Interface Control Documentation includes Interface Control Drawings, Interface Requirements Specifications, and other documentation that depicts physical and functional interfaces of related or co-functioning systems or components. ICD is the product of ICWGs or comparable integrated teams, and their purpose is to establish and maintain compatibility between interfacing systems or components.

Open Systems Interface Standards

To minimize the impact of unique interface designs, improve interoperability, maximize the use of commercial components, and improve the capacity for future upgrade, an open-systems approach should be a significant part of interface control planning. The open-systems approach involves selecting industry-recognized specifications and standards to define system internal and external interfaces. An open system is characterized by:

- Increased use of functional partitioning and modular design to enhance flexibility of component choices without impact on interfaces,

- Use of well-defined, widely used, non-proprietary interfaces or protocols based on standards developed or adopted by industry recognized standards institutions or professional societies, and

- Explicit provision for expansion or upgrading through the incorporation of additional or higher performance elements with minimal impact on the system.

DoD mandatory guidance for information technology standards is in the Joint Technical Architecture.

10.4 DATA MANAGEMENT

Data management documents and maintains the database reflecting system life cycle decisions, methods, feedback, metrics, and configuration control. It directly supports the configuration status accounting process. Data Management governs and controls the selection, generation, preparation, acquisition, and use of data imposed on contractors.

Data Required By Contract

Data is defined as recorded information, regardless of form or characteristic, and includes all the administrative, management, financial, scientific, engineering, and logistics information and documentation required for delivery from the contractor. Contractually required data is classified as one of three types:

- Type I: Technical data

- Type II: Non-technical data

- Type III: One-time use data (technical or non-technical)

Data is acquired for two basic purposes:

- Information feedback from the contractor for program management control, and

- Decision making information needed to manage, operate, and support the system (e.g., specifications, technical manuals, engineering drawings, etc.).

Data analysis and management is expensive and time consuming. Present DoD philosophy requires that the contractor manage and maintain significant portions of the technical data, including the Technical Data Package (TDP). Note that this does *not* mean the government isn't paying for its development or shouldn't receive a copy for post-delivery use. Minimize the TDP cost by requesting the contractor's format (for example, accepting the same drawings they use for production), and asking only for details on items developed with government funds.

Data Call for Government Contracts

As part of the development of an Invitation for Bid or Request for Proposals, the program office issues a letter that describes the planned procurement and asks integrated team leaders and effected functional managers to identify and justify their data requirements for that contract. A description of each data item needed is then developed by the affected teams or functional offices, and reviewed by the program office. Data Item Descriptions, located in the Acquisition Management Systems Data List (AMSDL) (see Chapter 8) can be used for guidance in developing these descriptions.

Concurrent with the DoD policy on specifications and standards, there is a trend to avoid use of standard Data Item Descriptions on contracts, and specify the data item with a unique tailored data description referenced in the Contract Data Requirements List.

10.5 SUMMARY POINTS

- Configuration management is essential to control the system design throughout the life cycle.

- Use of integrated teams in an IPPD environment is necessary for disciplined configuration management of complex systems.

- Technical data management is essential to trace decisions and changes and to document designs, processes and procedures.

- Interface management is essential to ensure that system elements are compatible in terms of form, fit, and function.

- Three configuration baselines are managed:
    – Functional (System level)
    – Allocated (Design To)
    – Product (Build To/As Built)

Configuration management is a shared responsibility between the government and the contractor. Contract manager (CM) key elements are Identification, Control, Status Accounting, and Audits.

CHAPTER 11

TECHNICAL REVIEWS AND AUDITS

11.1 PROGRESS MEASUREMENT

The Systems Engineer measures design progress and maturity by assessing its development at key event-driven points in the development schedule. The design is compared to pre-established exit criteria for the particular event to determine if the appropriate level of maturity has been achieved. These key events are generally known as Technical Reviews and Audits.

A system in development proceeds through a sequence of stages as it proceeds from concept to finished product. These are referred to as "levels of development." Technical Reviews are done after each level of development to check design maturity, review technical risk, and determines whether to proceed to the next level of development. Technical Reviews reduce program risk and ease the transition to production by:

- Assessing the maturity of the design/development effort,

- Clarifying design requirements,

- Challenging the design and related processes,

- Checking proposed design configuration against technical requirements, customer needs, and system requirements,

- Evaluating the system configuration at different stages,

- Providing a forum for communication, coordination, and integration across all disciplines and IPTs,

- Establishing a common configuration baseline from which to proceed to the next level of design, and

- Recording design decision rationale in the decision database.

Formal technical reviews are preceded by a series of technical interchange meetings where issues, problems and concerns are surfaced and addressed. The formal technical review is NOT the place for problem solving, but to verify problem solving has been done; it is a process rather than an event!

Planning

Planning for Technical Reviews must be extensive and up-front-and-early. Important considerations for planning include the following:

- Timely and effective attention and visibility into the activities preparing for the review,

- Identification and allocation of resources necessary to accomplish the total review effort,

- Tailoring consistent with program risk levels,

- Scheduling consistent with availability of appropriate data,

- Establishing event-driven entry and exit criteria,

- Where appropriate, conduct of incremental reviews,

- Implementation by IPTs,

- Review of all system functions, and

- Confirmation that all system elements are integrated and balanced.

The maturity of enabling products are reviewed with their associated end product. Reviews should consider the testability, producibility, training, and supportability for the system, subsystem or configuration item being addressed.

The depth of the review is a function of the complexity of the system, subsystem, or configuration item being reviewed. Where design is pushing state-of-the-art technology the review will require a greater depth than if it is for a commercial off-the-shelf item. Items, which are complex or an application of new technology, will require a more detailed scrutiny.

Planning Tip: Develop a check list of pre-review, review, and post-review activities required. Develop check lists for exit criteria and required level of detail in design documentation. Include key questions to be answered and what information must be available to facilitate the review process. Figure 11-1 shows the review process with key activities identified.

11.2 TECHNICAL REVIEWS

Technical reviews are conducted at both the system level and at lower levels (e.g., sub-system). This discussion will focus on the primary system-level reviews. Lower-level reviews may be thought of as events that support and prepare for the system-level events. The names used in reference to

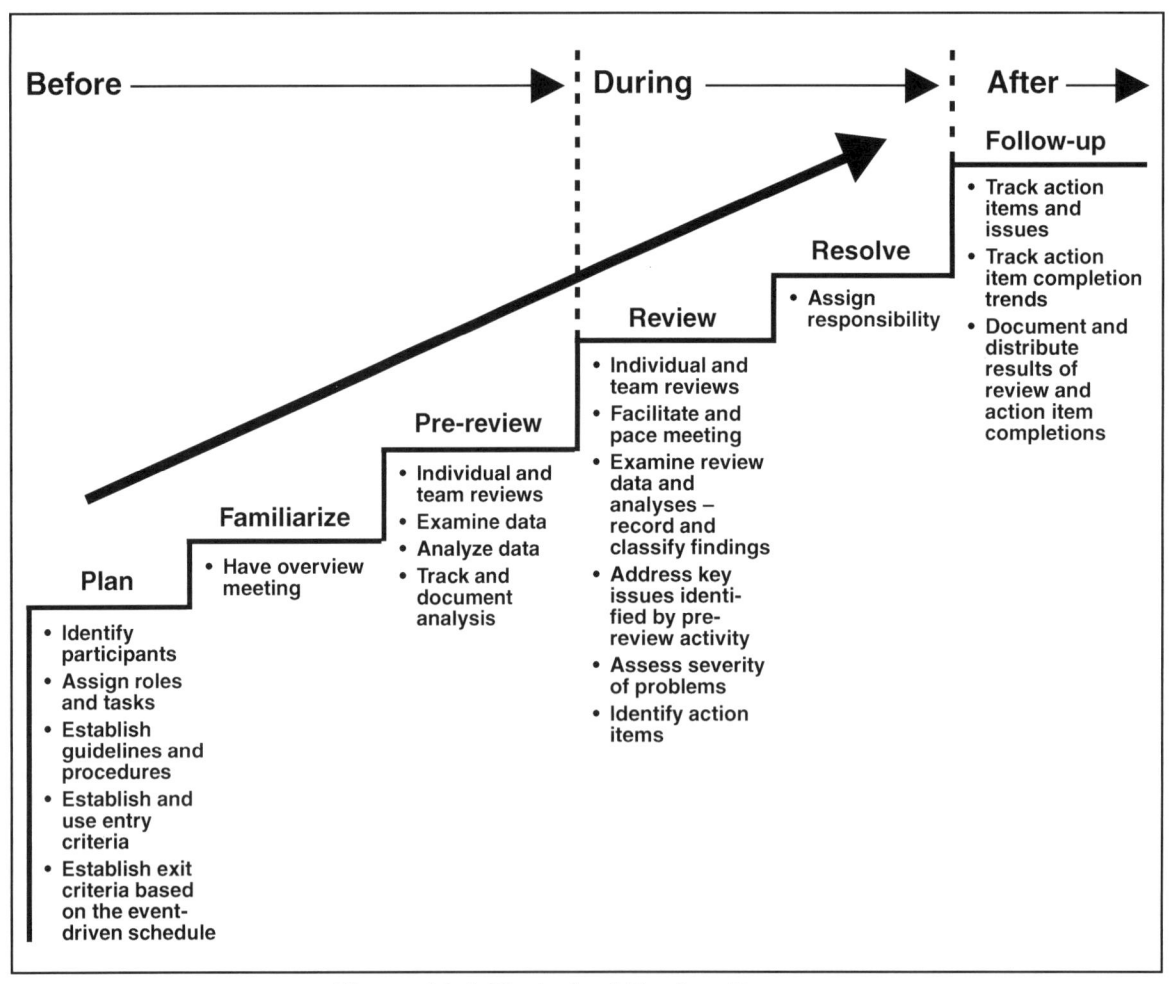

Figure 11-1. Technical Review Process

reviews is unimportant; however, it is important that reviews be held at appropriate points in program development and that both the contractor and government have common expectations regarding the content and outcomes.

Conducting Reviews

Reviews are event-driven, meaning that they are to be conducted when the progress of the product under development merits review. Forcing a review (simply based on the fact that a schedule developed earlier) projected the review at a point in time will jeopardize the review's legitimacy. Do the work ahead of the review event. Use the review event as a confirmation of completed effort. The data necessary to determine if the exit criteria are satisfied should be distributed, analyzed, and analysis coordinated prior to the review. The type of information needed for a technical review would include: specifications, drawings, manuals, schedules, design and test data, trade studies, risk analysis, effectiveness analyses, mock-ups, breadboards, in-process and finished hardware, test methods, technical plans (Manufacturing, Test, Support, Training), and trend (metrics) data. Reviews should be brief and follow a prepared agenda based on the pre-review analysis and assessment of where attention is needed.

Only designated participants should personally attend. These individuals should be those that were involved in the preparatory work for the review and members of the IPTs responsible for meeting the event exit criteria. Participants should include representation from all appropriate government activities, contractor, subcontractors, vendors and suppliers.

A review is the confirmation of a process. New items should not come up at the review. If significant items do emerge, it's a clear sign the review is

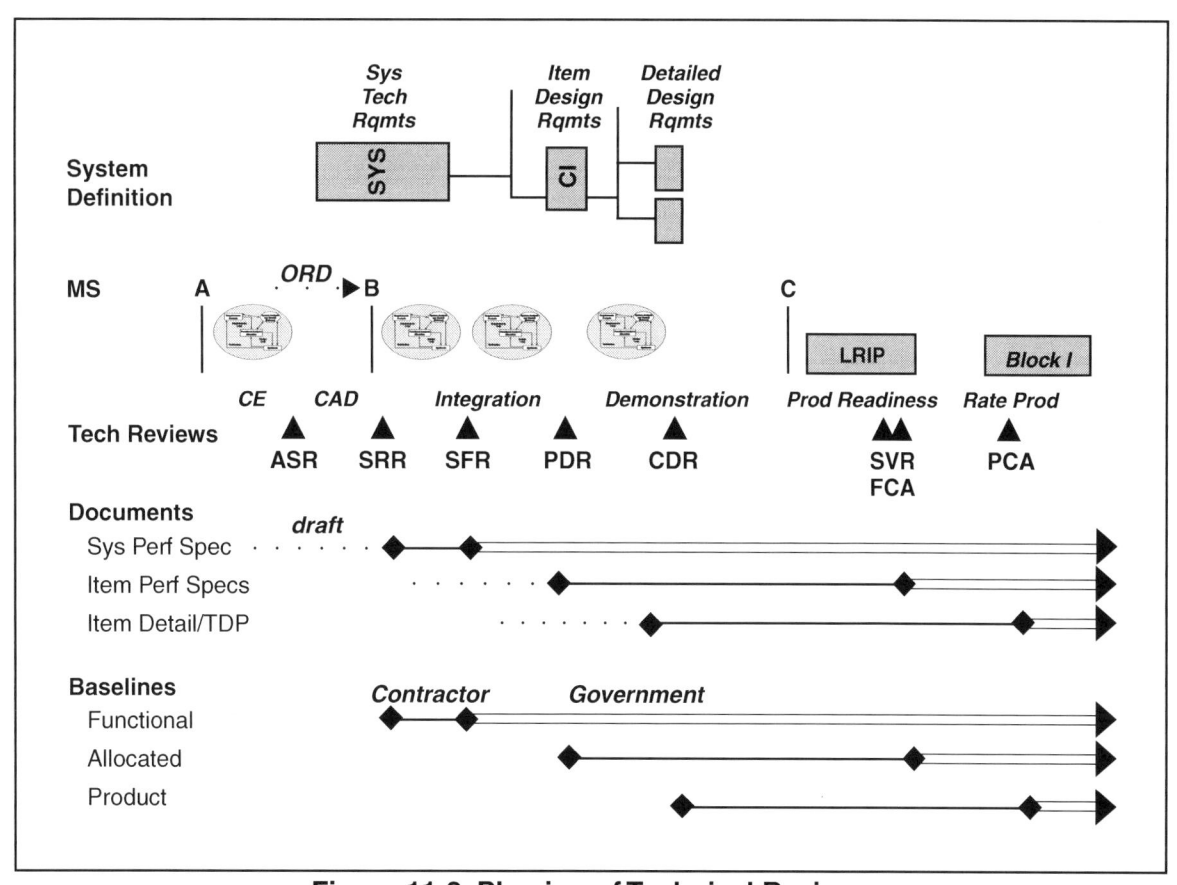

Figure 11-2. Phasing of Technical Reviews

being held prematurely, and project risk has just increased significantly. A poorly orchestrated and performed technical review is a significant indicator of management problems.

Action items resulting from the review are documented and tracked. These items, identified by specific nomenclature and due dates, are prepared and distributed as soon as possible after the review. The action taken is tracked and results distributed as items are completed.

Phasing of Technical Reviews

As a system progresses through design and development, it typically passes from a given level of development to another, more advanced level of development. For example, a typical system will pass from a stage where only the requirements are known, to another stage where a conceptual solution has been defined. Or it may pass from a stage where the design requirements for the primary subsystems are formalized, to a stage where the physical design solutions for those requirements are defined. (See Figure 11-2.)

These stages are the "levels of development" referred to in this chapter. System-level technical reviews are generally timed to correspond to the transition from one level of development to another. The technical review is the event at which the technical manager verifies that the technical maturity of the system or item under review is sufficient to justify passage into the subsequent phase of development, with the concomitant commitment of resources required.

As the system or product progresses through development, the focus of technical assessment takes different forms. Early in the process, the primary focus is on defining the requirements on which subsequent design and development activities will be based. Similarly, technical reviews conducted during the early stages of development are almost always focused on ensuring that the top-level concepts and system definitions reflect the requirements of the user. Once system-level definition is complete, the focus turns to design at sub-system levels and below. Technical reviews during these stages are typically design reviews that establish design requirements and then

Requirements Reviews
- Alternative System Review
- System Requirements Review

Design Reviews
- System Functional Review
- Preliminary Design Review (includes System Software Specification Review)
- Critical Design Review

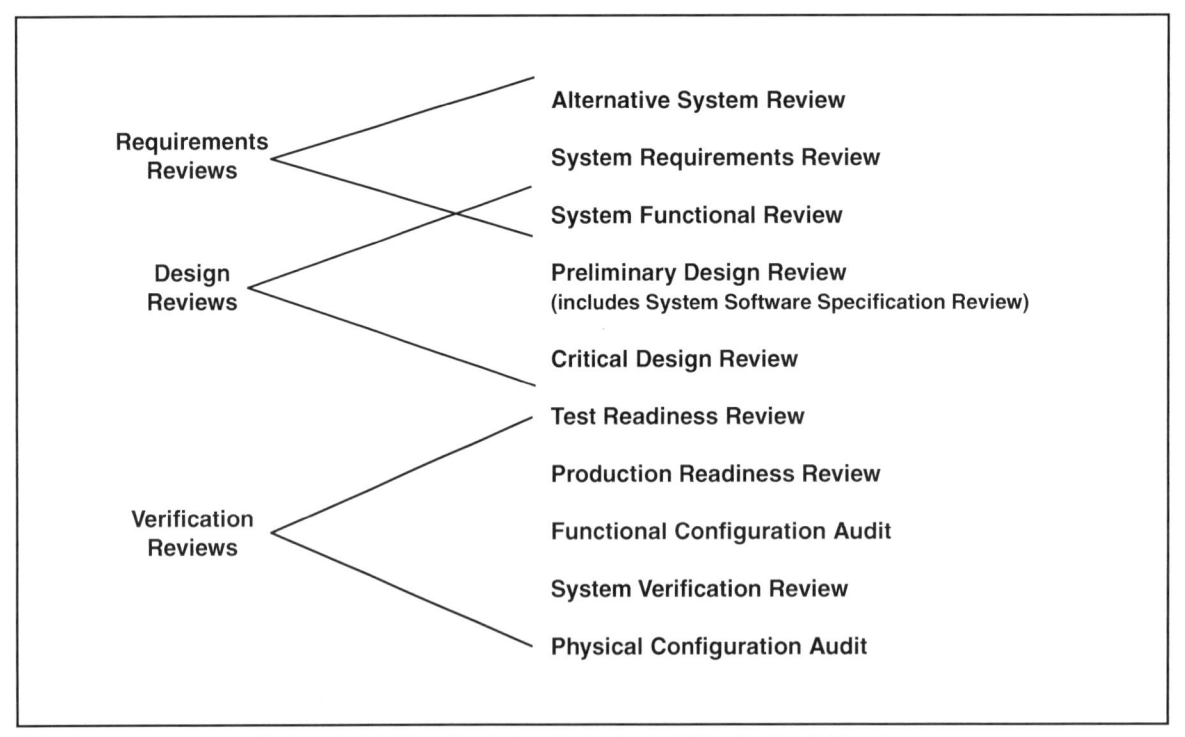

Verification Reviews
- Test Readiness Review
- Production Readiness Review
- Functional Configuration Audit
- System Verification Review
- Physical Configuration Audit

Figure 11-3. Typical System-Level Technical Reviews

verify that physical solutions are consistent with those requirements. In the final stages of development, technical reviews and audits are conducted to verify that the products produced meet the requirements on which the development is based. Figure 11-3 summarizes the typical schedule of system-level reviews by type and focus.

Another issue associated with technical reviews, as well as other key events normally associated with executing the systems engineering process, is when those events generally occur relative to the phases of the DoD acquisition life-cycle process. The timing of these events will vary somewhat from program to program, based upon the explicit and unique needs of the situation; however, Figure 11-4 shows a generalized concept of how the technical reviews normal to systems engineering might occur relative to the acquisition life-cycle phases.

Specific system-level technical reviews are known by many different names, and different engineering standards and documents often use different nomenclature when referring to the same review. The names used to refer to technical reviews are unimportant; however, it is important to have a grasp of the schedule of reviews that is normal to system development and to have an understanding of what is the focus and purpose of those reviews. The following paragraphs outline a schedule of reviews that is complete in terms of assessing technical progress from concept through production. The names used were chosen because they seemed to be descriptive of the focus of the activity. Of course, the array of reviews and the focus of individual reviews is to be tailored to the specific needs of the program under development, so not all programs should plan on conducting all of the following reviews.

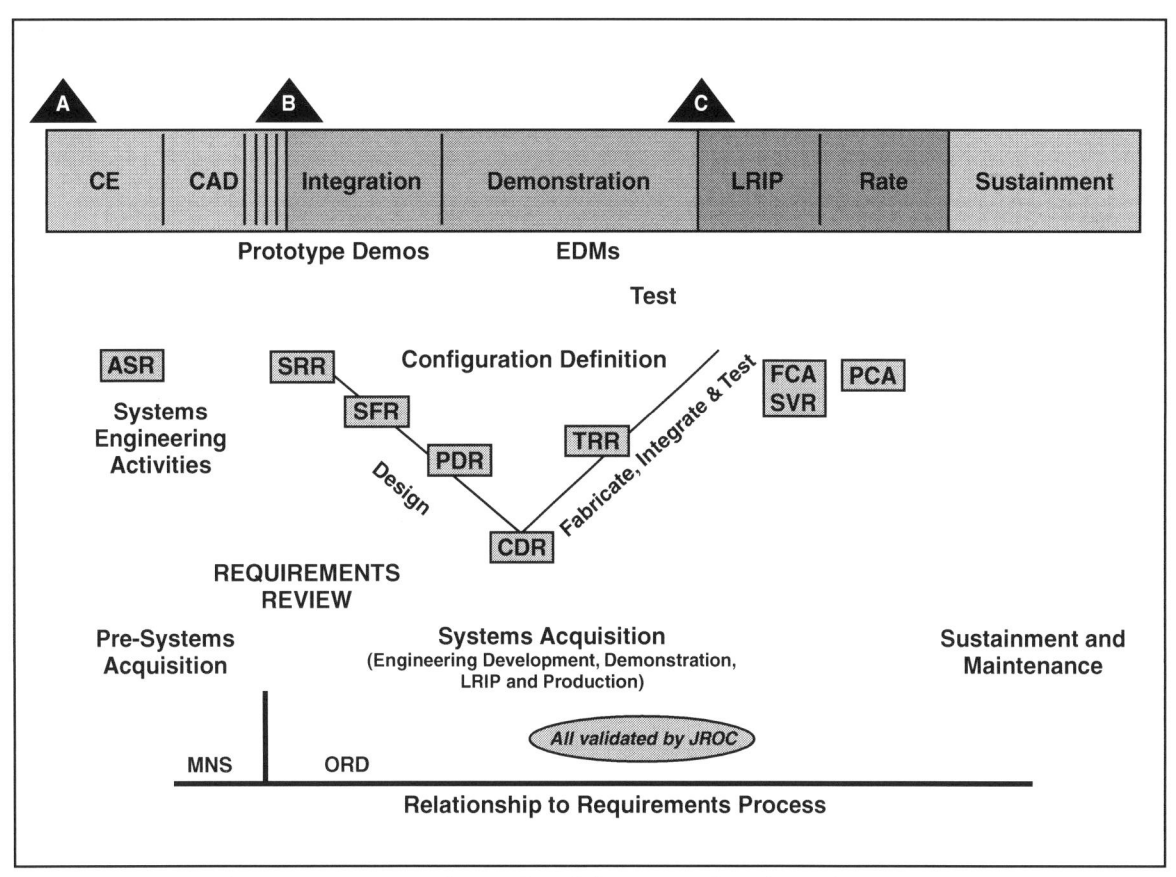

Figure 11-4. Relationship of Systems Engineering Events to Acquisition Life Cycle Phases

Alternative Systems Review (ASR)

After the concept studies are complete a preferred system concept is identified. The associated draft System Work Breakdown Structure, preliminary functional baseline, and draft system specification are reviewed to determine feasibility and risk. Technology dependencies are reviewed to ascertain the level of technology risk associated with the proposed concepts. This review is conducted late during the Concept Exploration stage of the Concept and Technology Development Phase of the acquisition process to verify that the preferred system concept:

- Provides a cost-effective, operationally-effective and suitable solution to identified needs,

- Meets established affordability criteria, and

- Can be developed to provide a timely solution to the need at an acceptable level of risk.

The findings of this review are a significant input to decision review conducted after Concept Exploration to determine where the system should enter in the life-cycle process to continue development. This determination is largely based on technology and system development maturity.

It is important to understand that the path of the system through the life-cycle process will be different for systems of different maturities. Consequently, the decision as whether or not to conduct the technical reviews that are briefly described in the following paragraphs is dependent on the extent of design and development required to bring the system to a level of maturity that justifies producing and fielding it.

System Requirements Review (SRR)

If a system architecture system must be developed and a top-down design elaborated, the system will pass through a number of well-defined levels of development, and that being the case, a well-planned schedule of technical reviews is imperative. The Component Advanced Development stage (the second stage of Concept and Technology Development in the revised acquisition life-cycle process) is the stage during which system-level architectures are defined and any necessary advanced development required to assess and control technical risk is conducted. As the system passes into the acquisition process, i.e., passes a Milestone B and enters System Development and Demonstration, it is appropriate to conduct a SRR. The SRR is intended to confirm that the user's requirements have been translated into system specific technical requirements, that critical technologies are identified and required technology demonstrations are planned, and that risks are well understood and mitigation plans are in place. The draft system specification is verified to reflect the operational requirements.

All relevant documentation should be reviewed, including:

- System Operational Requirements,

- Draft System Specification and any initial draft Performance Item Specifications,

- Functional Analysis (top level block diagrams),

- Feasibility Analysis (results of technology assessments and trade studies to justify system design approach),

- System Maintenance Concept,

- Significant system design criteria (reliability, maintainability, logistics requirements, etc.),

- System Engineering Planning,

- Test and Evaluation Master Plan,

- Draft top-level Technical Performance Measurement, and

- System design documentation (layout drawings, conceptual design drawings, selected supplier components data, etc.).

The SRR confirms that the system-level requirements are sufficiently well understood to permit the developer (contractor) to establish an initial system level functional baseline. Once that baseline is established, the effort begins to define the functional, performance, and physical attributes of the items below system level and to allocate them to the physical elements that will perform the functions.

System Functional Review (SFR)

The process of defining the items or elements below system level involves substantial engineering effort. This design activity is accompanied by analysis, trade studies, modeling and simulation, as well as continuous developmental testing to achieve an optimum definition of the major elements that make up the system, with associated functionality and performance requirements. This activity results in two major systems engineering products: the final version of the system performance specification and draft versions of the performance specifications, which describe the items below system level (item performance specifications). These documents, in turn, define the system functional baseline and the draft allocated baseline. As this activity is completed, the system has passed from the level of a concept to a well-defined system design, and, as such, it is appropriate to conduct another in the series of technical reviews.

The SFR will typically include the tasks listed below. Most importantly, the system technical description (Functional Baseline) must be approved as the governing technical requirement before proceeding to further technical development. This sets the stage for engineering design and development at the lower levels in the system architecture. The government, as the customer, will normally take control of and manage the system functional baseline following successful completion of the SFR.

The review should include assessment of the following items. More complete lists are found in standards and texts on the subject.

- Verification that the system specification reflects requirements that will meet user expectations,

- Functional Analysis and Allocation of requirements to items below system level,

- Draft Item Performance and some Item Detail Specifications,

- Design data defining the overall system,

- Verification that the risks associated with the system design are at acceptable levels for engineering development,

- Verification that the design selections have been optimized through appropriate trade study analyses,

- Supporting analyses, e.g., logistics, human systems integration, etc., and plans are identified and complete where appropriate,

- Technical Performance Measurement data and analysis, and

- Plans for evolutionary design and development are in place and that the system design is modular and open.

Following the SFR, work proceeds to complete the definition of the design of the items below system level, in terms of function, performance, interface requirements for each item. These definitions are typically captured in item performance specifications, sometimes referred to as prime item development specifications. As these documents are finalized, reviews will normally be held to verify that the design requirements at the item level reflect the set of requirements that will result in an acceptable detailed design, because all design work from the item level to the lowest level in the system will be based on the requirements agreed upon at the item level. The establishment of a set of final item-level design requirements represents the definition of the allocated baseline for the system. There are two primary reviews normally associated with this event: the Software Specification Review (SSR), and the Preliminary Design Review (PDR).

Software Specification Review (SSR)

As system design decisions are made, typically some functions are allocated to hardware items, while others are allocated to software. A separate specification is developed for software items to describe the functions, performance, interfaces and other information that will guide the design and development of software items. In preparation for the system-level PDR, the system software specification is reviewed prior to establishing the Allocated Baseline. The review includes:

- Review and evaluate the maturity of software requirements,

- Validation that the software requirements specification and the interface requirements specification reflect the system-level requirements allocated to software,

- Evaluation of computer hardware and software compatibility,

- Evaluation of human interfaces, controls, and displays

- Assurance that software-related risks have been identified and mitigation plans established,

- Validation that software designs are consistent with the Operations Concept Document,

- Plans for testing, and

- Review of preliminary manuals.

Preliminary Design Review (PDR)

Using the Functional Baseline, especially the System Specification, as a governing requirement, a preliminary design is expressed in terms of design requirements for subsystems and configuration items. This preliminary design sets forth the functions, performance, and interface requirements that will govern design of the items below system level. Following the PDR, this preliminary design (Allocated Baseline) will be put under formal configuration control [usually] by the contractor. The Item Performance Specifications, including the system software specification, which form the core of the Allocated Baseline, will be confirmed to represent a design that meets the System Specification.

This review is performed during the System Development and Demonstration phase. Reviews are held for configuration items (CIs), or groups of related CIs, prior to a system-level PDR. Item Performance Specifications are put under configuration control (Current DoD practice is for contractors to maintain configuration control over Item Performance Specifications, while the government exercises requirements control at the system level). At a minimum, the review should include assessment of the following items:

- Item Performance Specifications,

- Draft Item Detail, Process, and Material Specifications,

- Design data defining major subsystems, equipment, software, and other system elements,

- Analyses, reports, "ility" analyses, trade studies, logistics support analysis data, and design documentation,

- Technical Performance Measurement data and analysis,

- Engineering breadboards, laboratory models, test models, mockups, and prototypes used to support the design, and

- Supplier data describing specific components.

[Rough Rule of Thumb: ~15% of production drawings are released by PDR. This rule is anecdotal and only guidance relating to an "average" defense hardware program.]

Critical Design Review (CDR)

Before starting to build the production line there needs to be verification and formalization of the mutual understanding of the details of the item being produced. Performed during the System Development and Demonstration phase, this review evaluates the draft Production Baseline ("Build To" documentation) to determine if the system design documentation (Product Baseline, including Item Detail Specs, Material Specs, Process Specs) is satisfactory to start initial manufacturing. This review includes the evaluation of all CIs. It includes a series of reviews conducted for each hardware CI before release of design to fabrication, and each computer software CI before final coding and testing. Additionally, test plans are reviewed to assess if test efforts are developing sufficiently to indicate the Test Readiness Review will be successful. The approved detail design serves as the basis for final production planning and initiates the development of final software code.

[Rough Rule of Thumb: At CDR the design should be at least 85% complete. Many programs use drawing release as a metric for measuring design completion. This rule is anecdotal and only guidance relating to an "average" defense hardware program.]

Test Readiness Review (TRR)

Typically performed during the System Demonstration stage of the System Development and Demonstration phase (after CDR), the TRR assesses test objectives, procedures, and resources testing coordination. Originally developed as a software CI review, this review is increasingly applied to both hardware and software items. The TRR determines the completeness of test procedures and their compliance with test plans and descriptions. Completion coincides with the initiation of *formal* CI testing.

Production Readiness Reviews (PRR)

Performed incrementally during the System Development and Demonstration and during the Production Readiness stage of the Production and Deployment phase, this series of reviews is held to determine if production preparation for the system, subsystems, and configuration items is complete, comprehensive, and coordinated. PRRs are necessary to determine the readiness for production prior to executing a production go-ahead decision. They will formally examine the producibility of the production design, the control over the projected production processes, and adequacy of resources necessary to execute production. Manufacturing risk is evaluated in relationship to product and manufacturing process performance, cost, and schedule. These reviews support acquisition decisions to proceed to Low-Rate Initial Production (LRIP) or Full-Rate Production.

Functional Configuration Audit/ System Verification Review (FCA)/(SVR)

This series of audits and the consolidating SVR re-examines and verifies the customer's needs, and the relationship of these needs to the system and subsystem technical performance descriptions (Functional and Allocated Baselines). They determine if the system produced (including production representative prototypes or LRIP units) is capable of meeting the technical performance requirements established in the specifications, test plans, etc. The FCA verifies that all requirements established in the specifications, associated test plans, and related documents have been tested and that the item has passed the tests, or corrective action has been initiated. The technical assessments and decisions that are made in SVR will be presented to support the full-rate production go-ahead decision. Among the issues addressed:

- Readiness issues for continuing design, continuing verifications, production, training, deployment, operations, support, and disposal have been resolved,

- Verification is comprehensive and complete,

- Configuration audits, including completion of all change actions, have been completed for all CIs,

- Risk management planning has been updated for production,

- Systems Engineering planning is updated for production, and

- Critical achievements, success criteria and metrics have been established for production.

Physical Configuration Audit (PCA)

After full-rate production has been approved, follow-on independent verification (FOT&E) has identified the changes the user requires, and those changes have been corrected on the baseline documents and the production line, then it is time to assure that the product and the product baseline documentation are consistent. The PCA will formalize the Product Baseline, including specifications and the technical data package, so that future changes can only be made through full configuration management procedures. Fundamentally, the PCA verifies the product (as built) is consistent with the Technical Data Package which describes the Product Baseline. The final PCA confirms:

- The subsystem and CI PCAs have been successfully completed,

- The integrated decision database is valid and represents the product,

- All items have been baselined,

- Changes to previous baselines have been completed,

- Testing deficiencies have been resolved and appropriate changes implemented, and

- System processes are current and can be executed.

The PCA is a configuration management activity and is conducted following procedures established in the Configuration Management Plan.

11.3 TAILORING

The reviews described above are based on a complex system development project requiring significant technical evaluation. There are also cases where system technical maturity is more advanced than normal for the phase, for example, where a previous program or an Advanced Technical Concept Demonstration (ACTD) has provided a significant level of technical development applicable to the current program. In some cases this will precipitate the merging or even elimination of acquisition phases. This does not justify elimination of the technical management activities grouped under the general heading of systems analysis and control, nor does it relieve the government program manager of the responsibility to see that these disciplines are enforced. It does, however, highlight the need for flexibility and tailoring to the specific needs of the program under development.

For example, a DoD acquisition strategy that proposes that a system proceed directly into the demonstration stage may skip a stage of the complete acquisition process, but it must not skip the formulation of an appropriate Functional Baseline and the equivalent of an SFR to support the development. Nor should it skip the formulation of the Allocated Baseline and the equivalent of a PDR, and the formulation of the Product Baseline and the equivalent of a CDR. Baselines must be developed sequentially because they document different levels of design requirements and must build on each other. However, the assessment of design and development maturity can be tailored as appropriate for the particular system. Tailored efforts still have to deal with the problem of determining when the design maturity should be assessed, and how these assessments will support the formulation and control of baselines, which document the design requirements as the system matures.

In tailoring efforts, be extremely careful determining the level of system complexity. The system integration effort, the development of a single advanced technology or complex sub-component, or the need for intensive software development may be sufficient to establish the total system as a complex project, even though it appears simple because most subsystems are simple or off-the-shelf.

11.4 SUMMARY POINTS

- Each level of product development is evaluated and progress is controlled by specification development (System, Item Performance, Item Detail, Process, and Material specifications) and technical reviews and audits (ASR, SRR, SDR, SSR, PDR, CDR, TRR, PRR, FCA, SVR, PCA).

- Technical reviews assess development maturity, risk, and cost/schedule effectiveness to determine readiness to proceed.

- Reviews must be planned, managed, and followed up to be effective as an analysis and control tool.

- As the system progresses through the development effort, the nature of design reviews and audits will parallel the technical effort. Initially they will focus on requirements and functions, and later become very product focused.

- After system level reviews establish the Functional Baseline, technical reviews tend to be subsystem and CI focused until late in development when the focus again turns to the system level to determine the system's readiness for production.

CHAPTER 12

TRADE STUDIES

12.1 MAKING CHOICES

Trade Studies are a formal decision making methodology used by integrated teams to make choices and resolve conflicts during the systems engineering process. Good trade study analyses demand the participation of the integrated team; otherwise, the solution reached may be based on unwarranted assumptions or may reflect the omission of important data.

Trade studies identify desirable and practical alternatives among requirements, technical objectives, design, program schedule, functional and performance requirements, and life-cycle costs are identified and conducted. Choices are then made using a defined set of criteria. Trade studies are defined, conducted, and documented at the various levels of the functional or physical architecture in enough detail to support decision making and lead to a balanced system solution. The level of detail of any trade study needs to be commensurate with cost, schedule, performance, and risk impacts.

Both formal and informal trade studies are conducted in any systems engineering activity. Formal trade studies tend to be those that will be used in formal decision forums, e.g., milestone decisions. These are typically well documented and become a part of the decision database normal to systems development. On the other hand, engineering choices at every level involve trade-offs and decisions that parallel the trade study process. Most of these less-formal studies are documented in summary detail only, but they are important in that they define the design as it evolves.

Systems Engineering Process and Trade Studies

Trade studies are required to support decisions throughout the systems engineering process. During requirements analysis, requirements are balanced against other requirements or constraints, including cost. Requirements analysis trade studies examine and analyze alternative performance and functional requirements to resolve conflicts and satisfy customer needs.

During functional analysis and allocation, functions are balanced with interface requirements, dictated equipment, functional partitioning, requirements flowdown, and configuration items designation considerations. Trade studies are conducted within and across functions to:

- Support functional analyses and allocation of performance requirements and design constraints,

- Define a preferred set of performance requirements satisfying identified functional interfaces,

- Determine performance requirements for lower-level functions when higher-level performance and functional requirements can not be readily resolved to the lower-level, and

- Evaluate alternative functional architectures.

During design synthesis, trade studies are used to evaluate alternative solutions to optimize cost, schedule, performance, and risk. Trade studies are conducted during synthesis to:

- Support decisions for new product and process developments versus non-developmental products and processes;

- Establish system, subsystem, and component configurations;

- Assist in selecting system concepts, designs, and solutions (including people, parts, and materials availability);

- Support materials selection and make-or-buy, process, rate, and location decisions;

- Examine proposed changes;

- Examine alternative technologies to satisfy functional or design requirements including alternatives for moderate- to high- risk technologies;

- Evaluate environmental and cost impacts of materials and processes;

- Evaluate alternative physical architectures to select preferred products and processes; and

- Select standard components, techniques, services, and facilities that reduce system life-cycle cost and meet system effectiveness requirements.

During early program phases, for example, during Concept Exploration and functional baseline development, trade studies are used to examine alternative system-level concepts and scenarios to help establish the system configuration. During later phases, trade studies are used to examine lower-level system segments, subsystems, and end items to assist in selecting component part designs. Performance, cost, safety, reliability, risk, and other effectiveness measures must be traded against each other and against physical characteristics.

12.2 TRADE STUDY BASICS

Trade studies (trade-off analyses) are processes that examine viable alternatives to determine which is preferred. It is important that there be criteria established that are acceptable to all members of the integrated team as a basis for a decision. In addition, there must be an agreed-upon approach to measuring alternatives against the criteria. If these principles are followed, the trade study should produce decisions that are rational, objective, and repeatable. Finally, trade study results must be such that they can be easily communicated to customers and decision makers. If the results of a trade study are too complex to communicate with ease, it is unlikely that the process will result in timely decisions.

Trade Study Process

As shown by Figure 12-1, the process of trade-off analysis consists of defining the problem, bounding the problem, establishing a trade-off methodology (to include the establishment of decision criteria), selecting alternative solutions, determining the key characteristics of each alternative, evaluating the alternatives, and choosing a solution:

- Defining the problem entails developing a problem statement including any constraints. Problem definition should be done with extreme care. *After all, if you don't have the right problem, you won't get the right answer.*

- Bounding and understanding the problem requires identification of system requirements that apply to the study.

- Conflicts between desired characteristics of the product or process being studied, and the limitations of available data. Available databases should be identified that can provide relevant, historical "actual" information to support evaluation decisions.

- Establishing the methodology includes choosing the mathematical method of comparison, developing and quantifying the criteria used for comparison, and determining weighting factors (if any). Use of appropriate models and methodology will dictate the rationality, objectivity, and repeatability of the study. Experience has shown that this step can be easily abused

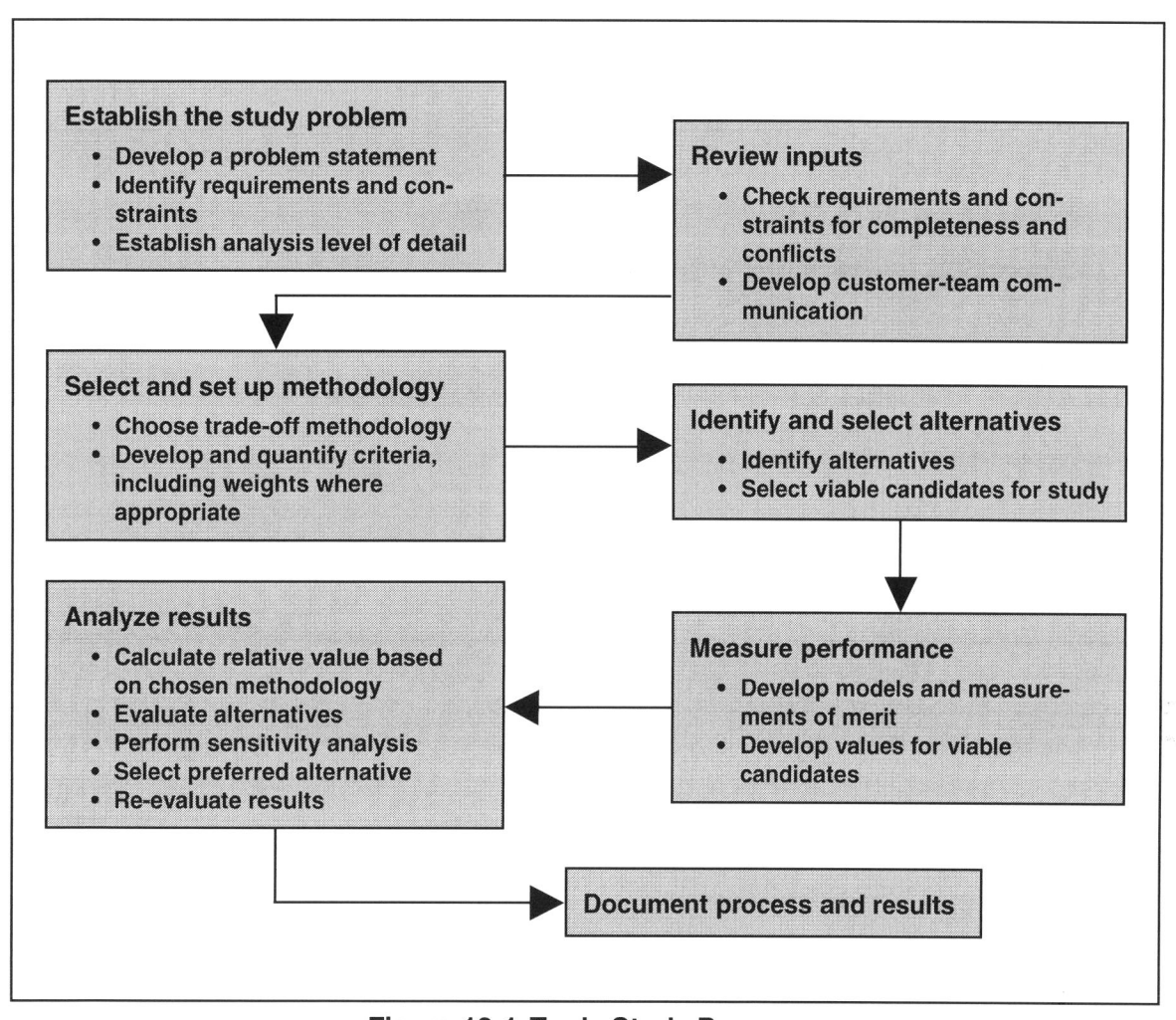

Figure 12-1. Trade Study Process

through both ignorance and design. To the extent possible the chosen methodology should compare alternatives based on true value to the customer and developer. Trade-off relationships should be relevant and rational. Choice of utility or weights should answer the question, "what is the actual value of the increased performance, based on what rationale?"

- Selecting alternative solutions requires identification of all the potential ways of solving the problem and selecting those that appear viable. The number of alternatives can drive the cost of analysis, so alternatives should normally be limited to clearly viable choices.

- Determining the key characteristics entails deriving the data required by the study methodology for each alternative.

- Evaluating the alternatives is the analysis part of the study. It includes the development of a trade-off matrix to compare the alternatives, performance of a sensitivity analysis, selection of a preferred alternative, and a re-evaluation (sanity check) of the alternatives and the study process. Since weighting factors and some "quantified" data can have arbitrary aspects, the sensitivity analysis is crucial. If the solution can be changed with relatively minor changes in data input, the study is probably invalid, and the methodology should be reviewed and revised. After the above tasks are complete, a solution is chosen, documented, and recorded in the database.

Cost Effectiveness Analyses

Cost effectiveness analyses are a special case trade study that compares system or component performance to its cost. These analyses help determine affordability and relative values of alternate solutions. Specifically, they are used to:

- Support identification of affordable, cost optimized mission and performance requirements,

- Support the allocation of performance to an optimum functional structure,

- Provide criteria for the selection of alternative solutions,

- Provide analytic confirmation that designs satisfy customer requirements within cost constraints, and

- Support product and process verification.

12.3 SUMMARY POINTS

- The purpose of trade studies is to make better and more informed decisions in selecting best alternative solutions.

- Initial trade studies focus on alternative system concepts and requirements. Later studies assist in selecting component part designs.

- Cost effectiveness analyses provide assessments of alternative solution performance relative to cost.

SUPPLEMENT 12-A

UTILITY CURVE METHODOLOGY

The utility curve is a common methodology used in DoD and industry to perform trade-off analysis. In DoD it is widely used for cost effectiveness analysis and proposal evaluation.

Utility Curve

The method uses a utility curve, Figure 12-2, for each of the decision factors to normalize them to ease comparison. This method establishes the relative value of the factor as it increases from the minimum value of the range. The curve shows can show a constant value relationship (straight line), increasing value (concave curve), decreasing value (convex curve), or a stepped value.

Decision Matrix

Each of the decision factors will also have relative value between them. These relative values are used to establish weighting factors for each decision factor. The weighting factors prioritize the decision factors and allow direct comparison between them. A decision matrix, similar to Figure 12-3, is generated to evaluate the relative value of the alternative solutions. In the case of Figure 12-3 range is given a weight of 2.0, speed a weight of 1.0, and payload a weight of 2.5. The utility values for each of the decision factors are multiplied by the appropriate weight. The weighted values for each alternative solution are added to obtain a total score for each solution. The solution with the highest score becomes the preferred solution. For the transport analysis of Figure 12-3 the apparent preferred solution is System 3.

Sensitivity

Figure 12-3 also illustrates a problem with the utility curve method. Both the utility curve and

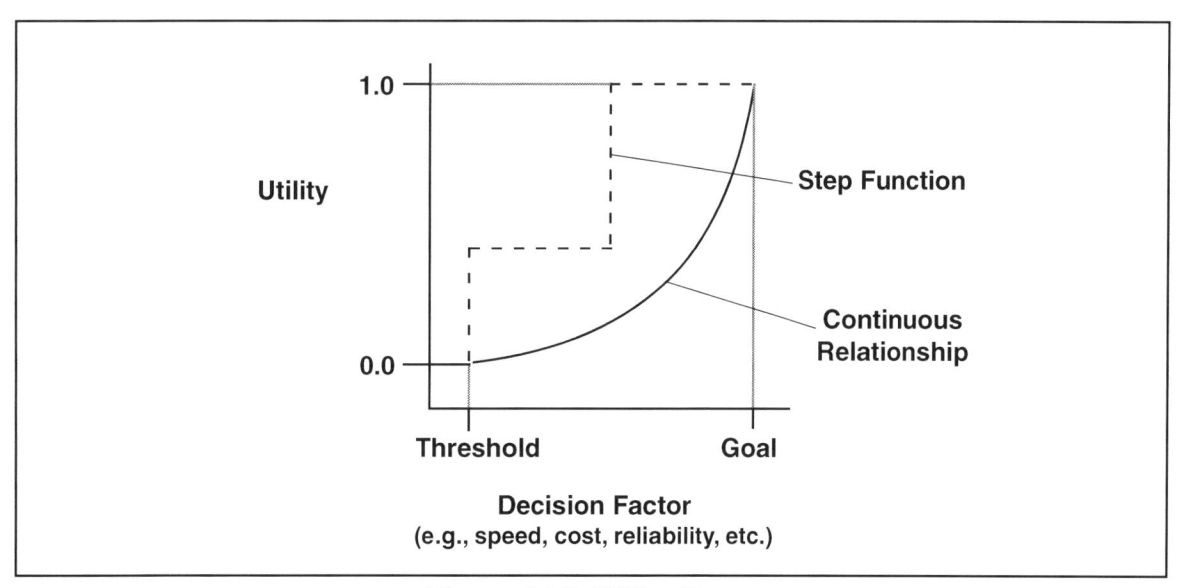

Figure 12-2. Utility Curve

weighting factors contain a degree of judgment that can vary between evaluators. Figure 12-3 shows three systems clustered around 3.8, indicating that a small variation in the utility curve or weighting factor could change the results. In the case of Figure 12-3, a sensitivity analysis should be performed to determine how solutions change as utility and weighting change. This will guide the evaluator in determining how to adjust evaluation criteria to eliminate the problem's sensitivity to small changes. In the case of Figure 12-3 the solution could be as simple as re-evaluating weighting factors to express better the true value to the customer. For example, if the value of range is considered to be less and payload worth more than originally stated, then System 4 may become a clear winner.

Notes

When developing or adjusting utility curves and weighting factors, communication with the customers and decision makers is essential. Most sensitivity problems are not as obvious as Figure 12-3. Sensitivity need not be apparent in the alternatives' total score. To ensure study viability, sensitivity analysis should always be done to examine the consequences of methodology choice. (Most decision support software provides a sensitivity analysis feature.)

Decision Factors / Alternatives	Range Wt. = 2.0 U	Range W	Speed Wt. = 1.0 U	Speed W	Payload Wt. = 2.5 U	Payload W	Weighted Total
Transport System 1	.8	1.6	.7	.7	.6	1.5	3.8
Transport System 2	.7	1.4	.9	.9	.4	1.0	3.3
Transport System 3	.6	1.2	.7	.7	.8	2.0	3.9
Transport System 4	.5	1.0	.5	.5	.9	2.25	3.75

Key: U = Utility value
W = Weighted value

Figure 12-3. Sample Decision Matrix

CHAPTER 13

MODELING AND SIMULATION

13.1 INTRODUCTION

A model is a physical, mathematical, or logical representation of a system entity, phenomenon, or process. A simulation is the implementation of a model over time. A simulation brings a model to life and shows how a particular object or phenomenon will behave. It is useful for testing, analysis or training where real-world systems or concepts can be represented by a model.

Modeling and simulation (M&S) provides virtual duplication of products and processes, and represents those products or processes in readily available and operationally valid environments. Use of models and simulations can reduce the cost and risk of life cycle activities. As shown by Figure 13-1, the advantages are significant throughout the life cycle.

Modeling, Simulation, and Acquisition

Modeling and simulation has become a very important tool across all acquisition-cycle phases and all applications: requirements definition; program management; design and engineering;

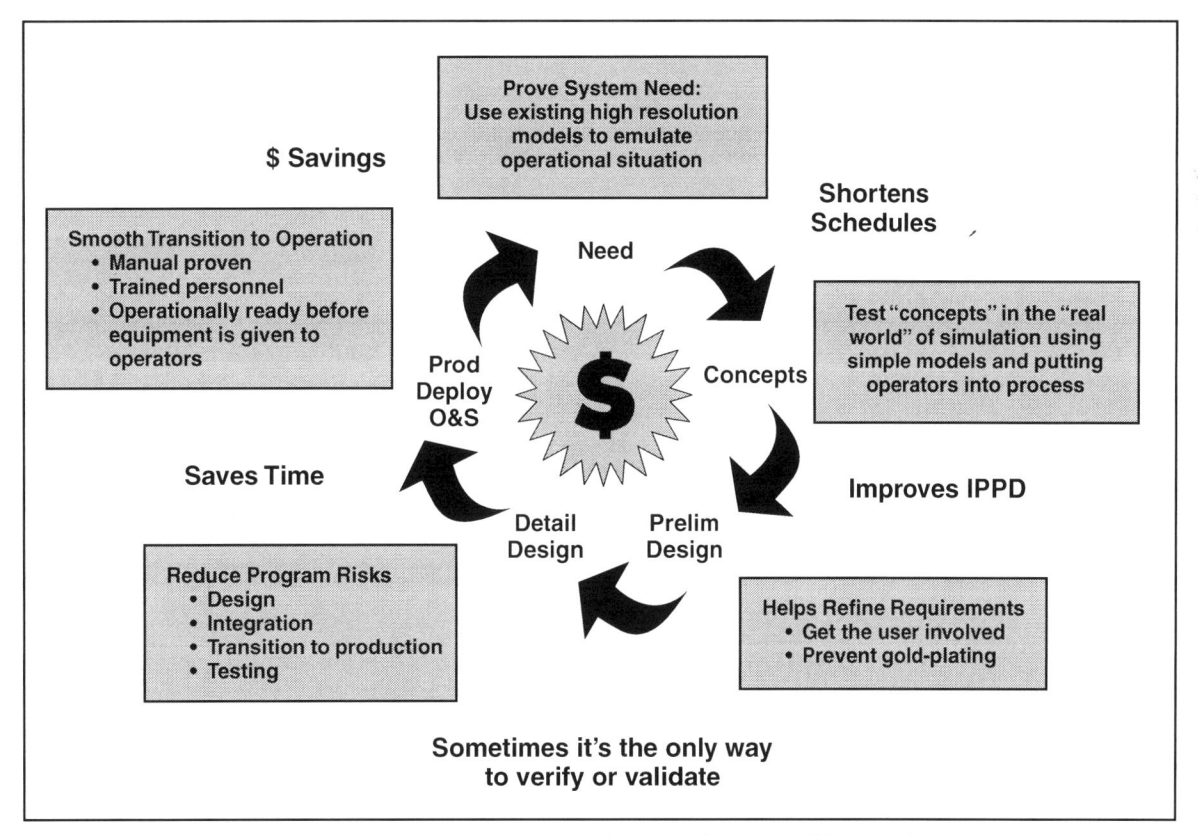

Figure 13-1. Advantages of Modeling and Simulation

efficient test planning; result prediction; supplement to actual test and evaluation; manufacturing; and logistics support. With so many opportunities to use M&S, its four major benefits; cost savings, accelerated schedule, improved product quality and cost avoidance can be achieved in any system development when appropriately applied. DoD and industry around the world have recognized these opportunities, and many are taking advantage of the increasing capabilities of computer and information technology. M&S is now capable of prototyping full systems, networks, interconnecting multiple systems and their simulators so that simulation technology is moving in every direction conceivable.

13.2 CLASSES OF SIMULATIONS

The three classes of models and simulations are virtual, constructive, and live:

- **Virtual** simulations represent systems both physically and electronically. Examples are aircraft trainers, the Navy's Battle Force Tactical Trainer, Close Combat Tactical Trainer, and built-in training.

- **Constructive** simulations represent a system and its employment. They include computer models, analytic tools, mockups, IDEF, Flow Diagrams, and Computer-Aided Design/ Manufacturing (CAD/CAM).

- **Live** simulations are simulated operations with real operators and real equipment. Examples are fire drills, operational tests, and initial production run with soft tooling.

Virtual Simulation

Virtual simulations put the human-in-the-loop. The operator's physical interface with the system is duplicated, and the simulated system is made to perform as if it were the real system. The operator is subjected to an environment that looks, feels, and behaves like the real thing. The more advanced version of this is the virtual prototype, which allows the individual to interface with a virtual mockup operating in a realistic computer-generated environment. A virtual prototype is a computer-based simulation of a system or subsystem with a degree of functional realism that is comparable to that of a physical prototype.

Constructive Simulations

The purpose of systems engineering is to develop descriptions of system solutions. Accordingly, constructive simulations are important products in all key system engineering tasks and activities. Of special interest to the systems engineer are Computer-Aided Engineering (CAE) tools. Computer-aided tools can allow more in-depth and complete analysis of system requirements early in design. They can provide improved communication because data can be disseminated rapidly to several individuals concurrently, and because design changes can be incorporated and distributed expeditiously. Key computer-aided engineering tools are CAD, CAE, CAM, Continuous Acquisition and Life Cycle Support, and Computer-Aided Systems Engineering:

Computer-Aided Design (CAD). CAD tools are used to describe the product electronically to facilitate and support design decisions. It can model diverse aspects of the system such as how components can be laid out on electrical/electronic circuit boards, how piping or conduit is routed, or how diagnostics will be performed. It is used to lay out systems or components for sizing, positioning, and space allocating using two- or three-dimensional displays. It uses three-dimensional "solid" models to ensure that assemblies, surfaces, intersections, interfaces, etc., are clearly defined. Most CAD tools automatically generate isometric and exploded views of detailed dimensional and assembly drawings, and determine component surface areas, volumes, weights, moments of inertia, centers of gravity, etc. Additionally, many CAD tools can develop three-dimensional models of facilities, operator consoles, maintenance workstations, etc., for evaluating man-machine interfaces. CAD tools are available in numerous varieties, reflecting different degrees of capabilities, fidelity, and cost. The commercial CAD/CAM product, Computer-Aided Three-Dimensional

Interactive Application (CATIA), was used to develop the Boeing 777, and is a good example of current state-of-the-art CAD.

Computer-Aided Engineering (CAE). CAE provides automation of requirements and performance analyses in support of trade studies. It normally would automate technical analyses such as stress, thermodynamic, acoustic, vibration, or heat transfer analysis. Additionally, it can provide automated processes for functional analyses such as fault isolation and testing, failure mode, and safety analyses. CAE can also provide automation of life-cycle-oriented analysis necessary to support the design. Maintainability, producibility, human factor, logistics support, and value/cost analyses are available with CAE tools.

Computer-Aided Manufacturing (CAM). CAM tools are generally designed to provide automated support to both production process planning and to the project management process. Process planning attributes of CAM include establishing Numerical Control parameters, controlling machine tools using pre-coded instructions, programming robotic machinery, handling material, and ordering replacement parts. The production management aspect of CAM provides management control over production-relevant data, uses historical actual costs to predict cost and plan activities, identifies schedule slips or slack on a daily basis, and tracks metrics relative to procurement, inventory, forecasting, scheduling, cost reporting, support, quality, maintenance, capacity, etc. A common example of a computer-based project planning and control tool is Manufacturing Resource Planning II (MRP II). Some CAM programs can accept data direct from a CAD program. With this type of tool, generally referred to as CAD/CAM, substantial CAM data is automatically generated by importing the CAD data directly into the CAM software.

Computer-Aided Systems Engineering (CASE). CASE tools provide automated support for the Systems Engineering and associated processes. CASE tools can provide automated support for integrating system engineering activities, performing the systems engineering tasks outlined in previous chapters, and performing the systems analysis and control activities. It provides technical management support and has a broader capability than either CAD or CAE. An increasing variety of CASE tools are available, as competition brings more products to market, and many of these support the commercial "best Systems Engineering practices."

Continuous Acquisition and Life Cycle Support (CALS). CALS relates to the application of computerized technology to plan and implement support functions. The emphasis is on information relating to maintenance, supply support, and associated functions. An important aspect of CALS is the importation of information developed during design and production. A key CALS function is to support the maintenance of the system configuration during the operation and support phase. In DoD, CALS supports activities of the logistics community rather than the specific program office, and transfer of data between the CAD or CAM programs to CALS has been problematic. As a result there is current emphasis on development of standards for compatible data exchange. Formats of import include: two- and three-dimensional models (CAD), ASCII formats (Technical Manuals), two-dimensional illustrations (Technical Manuals), and Engineering Drawing formats (Raster, Aperture cards). These formats will be employed in the Integrated Data Environment (IDE) that is mandated for use in DoD program offices.

Live Simulation

Live simulations are simulated operations of real systems using real people in realistic situations. The intent is to put the system, including its operators, through an operational scenario, where some conditions and environments are mimicked to provide a realistic operating situation. Examples of live simulations range from fleet exercises to fire drills.

Eventually live simulations must be performed to validate constructive and virtual simulations. However, live simulations are usually costly, and trade studies should be performed to support the balance of simulation types chosen for the program.

13.3 HARDWARE VERSUS SOFTWARE

Though current emphasis is on software M&S, the decision of whether to use hardware, software, or a combined approach is dependent on the complexity of the system, the flexibility needed for the simulation, the level of fidelity required, and the potential for reuse. Software capabilities are increasing, making software solutions cost effective for large complex projects and repeated processes. Hardware methods are particularly useful for validation of software M&S, simple or one-time projects, and quick checks on changes of production systems. M&S methods will vary widely in cost. Analysis of the cost-versus-benefits of potential M&S methods should be performed to support planning decisions.

13.4 VERIFICATION, VALIDATION, AND ACCREDITATION

How can you trust the model or simulation? Establish confidence in your model or simulation through formal verification, validation, and accreditation (VV&A). VV&A is usually identified with software, but the basic concept applies to hardware as well. Figure 13-2 shows the basic differences between the terms (VV&A).

More specifically:

- **Verification** is the process of determining that a model implementation accurately represents the developer's conceptual description and specifications that the model was designed to.

- **Validation** is the process of determining the manner and degree to which a model is an accurate representation of the real world from the perspective of the intended uses of the model, and of establishing the level of confidence that should be placed on this assessment.

- **Accreditation** is the formal certification that a model or simulation is acceptable for use for a specific purpose. Accreditation is conferred by the organization best positioned to make the judgment that the model or simulation in question is acceptable. That organization may be an operational user, the program office, or a contractor, depending upon the purposes intended.

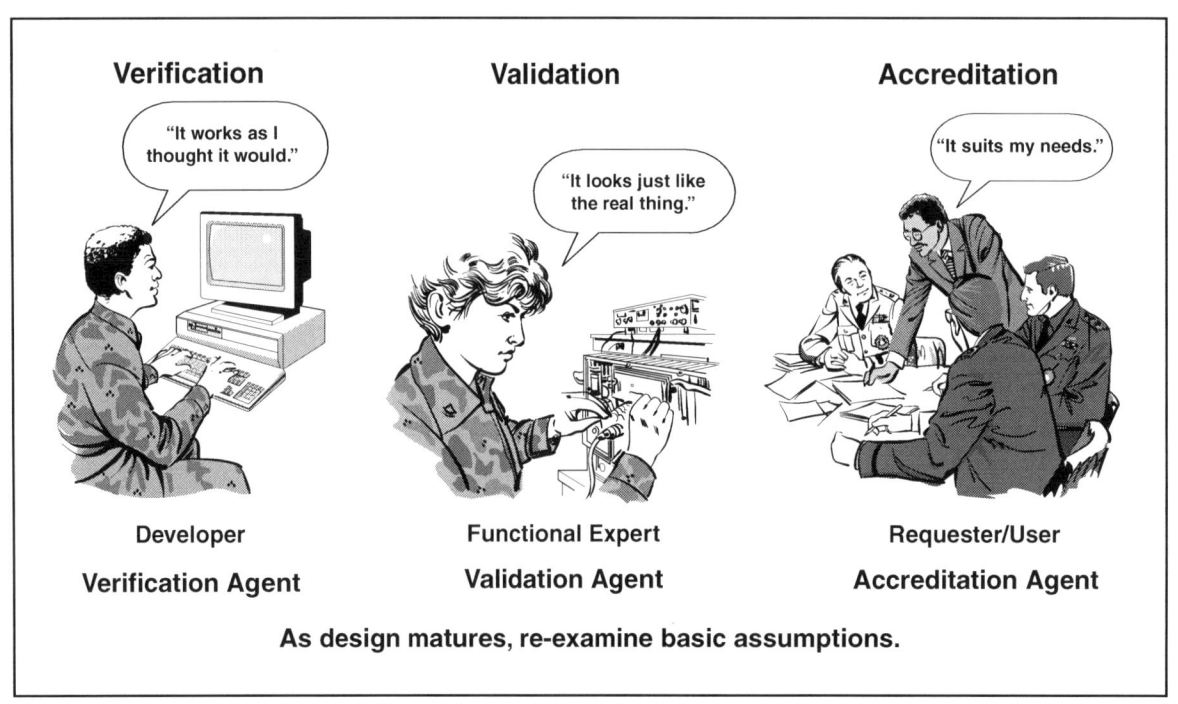

Figure 13-2. Verification, Validation, and Accreditation

VV&A is particularly necessary in cases where:

- Complex and critical interoperability is being represented,

- Reuse is intended,

- Safety of life is involved, and

- Significant resources are involved.

VV&A Currency

VV&A is applied at initial development and use. The VV&A process is required for all DoD simulations and should be redone whenever existing models and simulations undergo a major upgrade or modification. Additionally, whenever the model or simulation violates its documented methodology or inherent boundaries that were used to validate or verify by its different use, then VV&A must be redone. Accreditation, however, may remain valid for the specific application unless revoked by the Accreditation Agent, as long as its use or what it simulates doesn't change.

13.5 CONSIDERATIONS

There are a number of considerations that should enter into decisions regarding the acquisition and employment of modeling and simulation in defense acquisition management. Among these are such concerns as cost, fidelity, planning, balance, and integration.

Cost Versus Fidelity

Fidelity is the degree to which aspects of the real world are represented in M&S. It is the foundation for development of the model and subsequent VV&A. Cost effectiveness is a serious issue with simulation fidelity, because fidelity can be an aggressive cost driver. The correct balance between cost and fidelity should be the result of simulation need analysis. M&S designers and VV&A agents must decide when enough is enough. Fidelity needs can vary throughout the simulation. This variance should be identified by analysis and planned for.

Note of caution: Don't confuse the quality of the display with the quality of meeting simulation needs! An example of fidelity is a well-known flight simulator using a PC and simple joystick versus a full 6-degree of freedom fully-instrumented aircraft cockpit. Both have value at different stages of flight training, but obviously vary significantly in cost from thousands of dollars to millions. This cost difference is based on fidelity, or degree of real-world accuracy.

Planning

Planning should be an inherent part of M&S, and, therefore, it must be proactive, early, continuous, and regular. Early planning will help achieve balance and beneficial reuse and integration. With computer and simulation technologies evolving so rapidly, planning is a dynamic process. It must be a continuing process, and it is important that the appropriate simulation experts be involved to maximize the use of new capabilities. M&S activities should be a part of the integrated teaming and involve all responsible organizations. Integrated teams must develop their M&S plans and insert them into the overall planning process, including the TEMP, acquisition strategy, and any other program planning activity.

M&S planning should include:

- Identification of activities responsible for each VV&A element of each model or simulation, and

- Thorough VV&A estimates, formally agreed to by all activities involved in M&S, including T&E commitments from the developmental testers, operational testers, and separate VV&A agents.

Those responsible for the VV&A activities must be identified as a normal part of planning. Figure 13-2 shows the developer as the verification agent, the functional expert as the validation agent, and the user as the accreditation agent. In general this is appropriate for virtual simulations. However, the manufacturer of a constructive simulation would usually be expected to justify or warrantee their program's use for a particular application. The question of who should actually accomplish VV&A is one that is answered in planning. VV&A requirements should be specifically called out in tasking documents and contracts. When appropriate, VV&A should be part of the contractor's proposal, and negotiated prior to contract award.

Balance

Balance refers to the use of M&S across the phases of the product life cycle and across the spectrum of functional disciplines involved. The term may further refer to the use of hardware versus software, fidelity level, VV&A level, and even use versus non-use. Balance should always be based on cost effectiveness analysis. Cost effectiveness analyses should be comprehensive; that is, M&S should be properly considered for use in all parallel applications and across the complete life cycle of the system development and use.

Integration

Integration is obtained by designing a model or simulation to inter-operate with other models or simulations for the purpose of increased performance, cost benefit, or synergism. Multiple benefits or savings can be gained from increased synergism and use over time and across activities. Integration is achieved through reuse or upgrade of legacy programs used by the system, or of the proactive planning of integrated development of new simulations. In this case integration is accomplished through the planned utilization of models, simulations, or data for multiple times or applications over the system life cycle. The planned upgrade of M&S for evolving or parallel uses supports the application of open systems architecture to the system design. M&S efforts that are established to perform a specific function by a specific contractor, subcontractor, or government activity will tend to be sub-optimized. To achieve

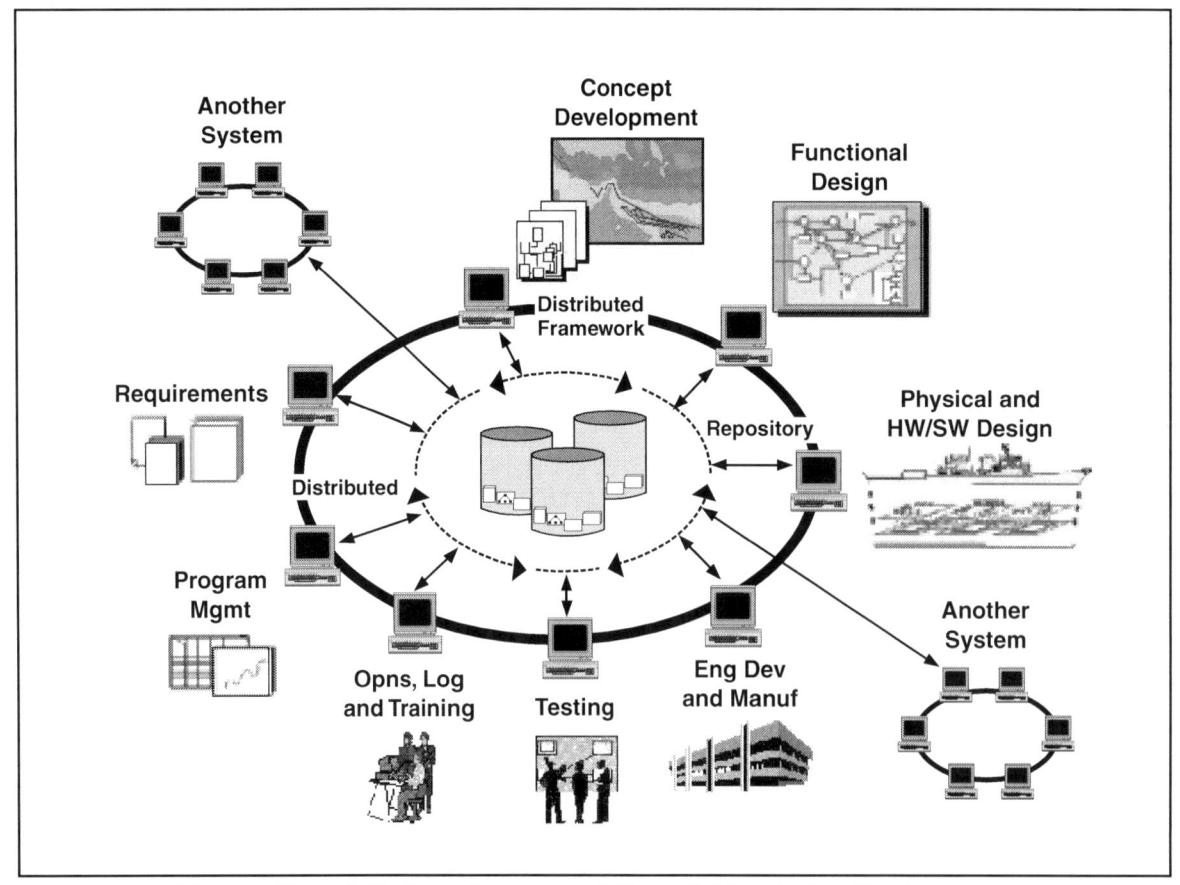

Figure 13-3. A Robust Integrated Use of Simulation Technology

integration M&S should be managed at least at the program office level.

The Future Direction

DoD, the Services, and their commands have strongly endorsed the use of M&S throughout the acquisition life cycle. The supporting simulation technology is also evolving as fast as computer technology changes, providing greater fidelity and flexibility. As more simulations are interconnected, the opportunities for further integration expand. M&S successes to date also accelerate its use. The current focus is to achieve open systems of simulations, so they can be plug-and-play across the spectrum of applications. From concept analysis through disposal analysis, programs may use hundreds of different simulations, simulators and model analysis tools. Figure 13-3 shows conceptually how an integrated program M&S would affect the functions of the acquisition process.

A formal DoD initiative, Simulation Based Acquisition (SBA), is currently underway. The SBA vision is to advance the implementation of M&S in the DoD acquisition process toward a robust, collaborative use of simulation technology that is integrated across acquisition phases and programs. The result will be programs that are much better integrated in an IPPD sense, and which are much more efficient in the use of time and dollars expended to meet the needs of operational users.

13.6 SUMMARY

- M&S provides virtual duplication of products and processes, and represent those products or processes in readily available and operationally valid environments.

- M&S should be applied throughout the system life cycle in support of systems engineering activities.

- The three classes of models and simulations are virtual, constructive, and live.

- Establish confidence in your model or simulation through formal VV&A.

- M&S planning should be an inherent part of Systems Engineering planning, and, therefore, pro-active, early, continuous, and regular.

- A more detailed discussion of the use and management of M&S in DoD acquisition is available in the DSMC publication *Systems Acquisition Manager's Guide for the Use of Models and Simulations*.

- An excellent second source is the DSMC publication, *Simulation Based Acquisition – A New Approach*. It surveys applications of increasing integration of simulation in current DoD programs and the resulting increasing benefits through greater integration.

CHAPTER 14

METRICS

14.1 METRICS IN MANAGEMENT

Metrics are measurements collected for the purpose of determining project progress and overall condition by observing the change of the measured quantity over time. Management of technical activities requires use of three basic types of metrics:

- Product metrics that track the development of the product,

- Earned Value which tracks conformance to the planned schedule and cost, and

- Management process metrics that track management activities.

Measurement, evaluation and control of metrics is accomplished through a system of periodic reporting must be planned, established, and monitored to assure metrics are properly measured, evaluated, and the resulting data disseminated.

Product Metrics

Product metrics are those that track key attributes of the design to observe progress toward meeting customer requirements. Product metrics reflect three basic types of requirements: operational performance, life-cycle suitability, and affordability. The key set of systems engineering metrics are the Technical Performance Measurements (TPM.) TPMs are product metrics that track design progress toward meeting customer performance requirements. They are closely associated with the system engineering process because they directly support traceability of operational needs to the design effort. TPMs are derived from Measures of Performance (MOPs) which reflect system requirements. MOPs are derived from Measures of Effectiveness (MOEs) which reflect operational performance requirements.

The term "metric" implies quantitatively measurable data. In design, the usefulness of metric data is greater if it can be measured at the configuration item level. For example, weight can be estimated at all levels of the WBS. Speed, though an extremely important operational parameter, cannot be allocated down through the WBS. It cannot be measured, except through analysis and simulation, until an integrated product is available. Since weight is an important factor in achieving speed objectives, and weight can be measured at various levels as the system is being developed, weight may be the better choice as a metric. It has a direct impact on speed, so it traces to the operational requirement, but, most importantly, it can be allocated throughout the WBS and progress toward achieving weight goals may then be tracked through development to production.

Measures of Effectiveness and Suitability

Measures of Effectiveness (MOEs) and Measures of Suitability (MOSs) are measures of operational effectiveness and suitability in terms of operational outcomes. They identify the most critical performance requirements to meet system-level mission objectives, and will reflect key operational needs in the operational requirements document.

Operational effectiveness is the overall degree of a system's capability to achieve mission success considering the total operational environment. For example, weapon system effectiveness would consider environmental factors such as operator organization, doctrine, and tactics; survivability; vulnerability; and threat characteristics. MOSs, on the other hand, would measure the extent to which the system integrates well into the operation environment and would consider such issues as supportability, human interface compatibility, and maintainability.

Measures of Performance

MOPs characterize physical or functional attributes relating to the execution of the mission or function. They quantify a technical or performance requirement directly derived from MOEs and MOSs. MOPs should relate to these measures such that a change in MOP can be related to a change in MOE or MOS. MOPs should also reflect key performance requirements in the system specification. MOPs are used to derive, develop, support, and document the performance requirements that will be the basis for design activities and process development. They also identify the critical technical parameters that will be tracked through TPMs.

Technical Performance Measurements

TPMs are derived directly from MOPs, and are selected as being critical from a periodic review and control standpoint. TPMs help assess design progress, assess compliance to requirements throughout the WBS, and assist in monitoring and tracking technical risk. They can identify the need for deficiency recovery, and provide information to support cost-performance sensitivity assessments. TPMs can include range, accuracy, weight, size, availability, power output, power required, process time, and other product characteristics that relate directly to the system operational requirements.

TPMs traceable to WBS elements are preferred, so elements within the system can be monitored as well as the system as a whole. However, some necessary TPMs will be limited to the system or subsystem level. For example, the specific fuel consumption of an engine would be a TPM necessary to track during the engine development, but it is not allocated throughout the WBS. It is reported as a single data item reflecting the performance of the engine as a whole. In this case the metric will indicate that the design approach is consistent with the required performance, but it may not be useful as an early warning device to indicate progress toward meeting the design goal. A more detailed discussion of TPMs is available as Supplement A to this chapter.

Example of Measures

MOE: The vehicle must be able to drive fully loaded from Washington, DC, to Tampa on one tank of fuel.

MOP: Vehicle range must be equal to or greater than 1,000 miles.

TPM: Fuel consumption, vehicle weight, tank size, drag, power train friction, etc.

Suitability Metrics

Tracking metrics relating to operational suitability and other life cycle concerns may be appropriate to monitor progress toward an integrated design. Operational suitability is the degree to which a system can be placed satisfactorily in field use considering availability, compatibility, transportability, interoperability, reliability, usage rates, maintainability, safety, human factors, documentation, training, manpower, supportability, logistics, and environmental impacts. These suitability parameters can generate product metrics that indicate progress toward an operationally suitable system. For example, factors that indicate the level of automation in the design would reflect progress toward achieving manpower quantity and quality requirements. TPMs and suitability product metrics commonly overlap. For example, Mean Time Between Failure (MBTF) can reflect both effectiveness or suitability requirements.

Suitability metrics would also include measurements that indicate improvement in the producibility, testability, degree of design simplicity, and design robustness. For example, tracking number of parts, number of like parts, and number of wearing parts provides indicators of producibility, maintainability, and design simplicity.

Product Affordability Metrics

Estimated unit production cost can be tracked during the design effort in a manner similar to the TPM approach, with each CI element reporting an estimate based on current design. These estimates are combined at higher WBS levels to provide subsystem and system cost estimates. This provides a running engineering estimate of unit production cost, tracking of conformance to Design-to-Cost (DTC) goals, and a method to isolate design problems relating to production costs.

Life cycle affordability can be tracked through factors that are significant in parametric life cycle cost calculations for the particular system. For example, two factors that reflect life cycle cost for most transport systems are fuel consumption and weight, both of which can be tracked as metrics.

Timing

Product metrics are tied directly to the design process. Planning for metric identification, reporting, and analysis is begun with initial planning in the concept exploration phase. The earliest systems engineering planning should define the management approach, identify performance or characteristics to be measured and tracked, forecast values for those performances or characteristics, determine when assessments will be done, and establish the objectives of assessment.

Implementation is begun with the development of the functional baseline. During this period, systems engineering planning will identify critical technical parameters, time phase planned profiles with tolerance bands and thresholds, reviews or audits or events dependent or critical for achievement of planned profiles, and the method of estimation. During the design effort, from functional to product baseline, the plan will be implemented and continually updated by the systems engineering process. To support implementation, contracts should include provision for contractors to provide measurement, analysis, and reporting. The need to track product metrics ends in the production phase, usually concurrent with the establishment of the product (as built) baseline.

DoD and Industry Policy on Product Metrics

Analysis and control activities shall include performance metrics to measure technical development and design, actual versus planned; and to measure [the extent to which systems meet requirements]. DoD 5000.2-R.

The performing activity establishes and implements TPM to evaluate the adequacy of evolving solutions to identify deficiencies impacting the ability of the system to satisfy a designated value for a technical parameter. EIA IS-632, Section 3.

The performing activity identifies the technical performance measures which are key indicators of system performance...should be limited to critical MOPs which, if not met put the project at cost, schedule, or performance risk. IEEE 1220, Section 6.

14.2 EARNED VALUE

Earned Value is a metric reporting system that uses cost-performance metrics to track the cost and schedule progress of system development against a projected baseline. It is a "big picture" approach and integrates concerns related to performance, cost, and schedule. Referring to Figure 14-1, if we think of the line labeled BCWP (budgeted cost of work performed) as the value that the contractor has "earned," then deviations from this baseline indicate problems in either cost or schedule. For example, if actual costs vary from budgeted costs, we have a cost variance; if work performed varies from work planned, we have a schedule variance. The projected performance is based on estimates of appropriate cost and schedule to perform the work required by each WBS element. When a variance occurs the system engineer can pinpoint WBS elements that have potential technical development problems. Combined with product metrics, earned value is a powerful technical management tool for detecting and understanding development problems.

Relationships exist between product metrics, the event schedule, the calendar schedule, and Earned

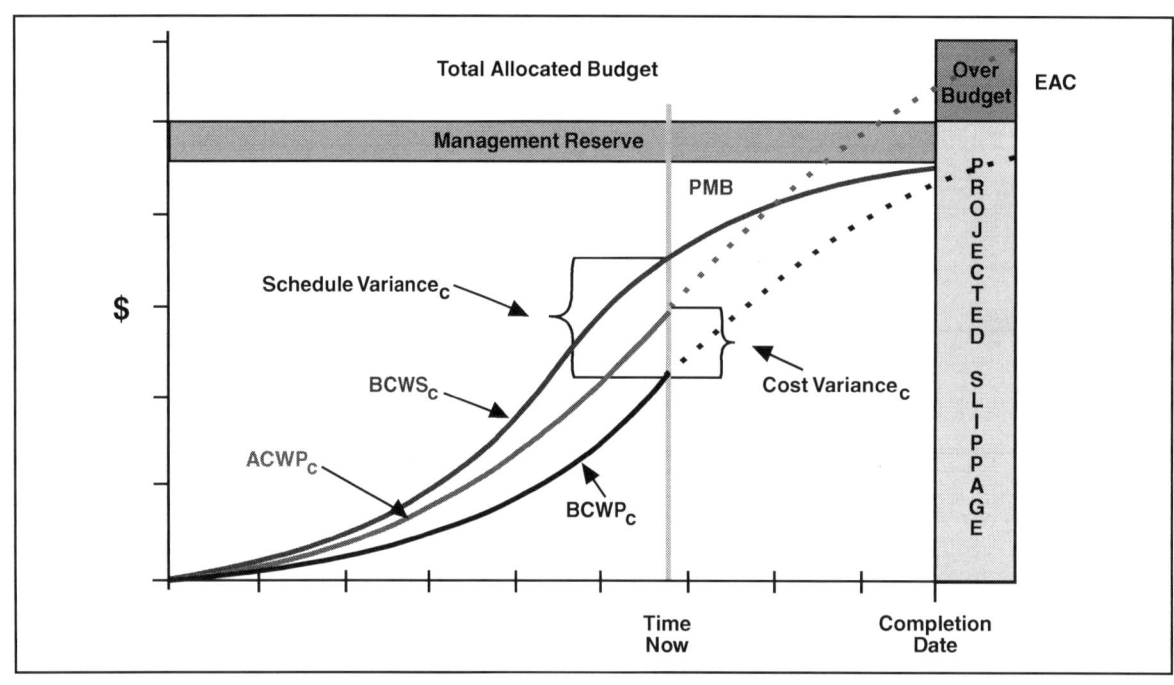

Figure 14-1. Earned Value Concept

Value:

- The Event Schedule includes tasks for each event/exit criteria that must be performed to meet key system requirements, which are directly related to product metrics.

- The Calendar (Detail) Schedule includes time frames established to meet those same product metric-related objectives (schedules).

- Earned Value includes cost/schedule impacts of not meeting those objectives, and, when correlated with product metrics, can identify emerging program and technical risk.

14.3 PROCESS METRICS

Management process metrics are measurements taken to track the process of developing, building, and introducing the system. They include a wide range of potential factors and selection is program unique. They measure such factors as availability of resources, activity time rates, items completed, completion rates, and customer or team satisfaction. Examples of these factors are: number of trained personnel onboard, average time to approve/disapprove ECPs, lines of code or drawings released, ECPs resolved per month, and team risk identification or feedback assessments. Selection of appropriate metrics should be done to track key management activities. Selection of these metrics is part of the systems engineering planning process.

How Much Metrics?

The choice of the amount and depth of metrics is a planning function that seeks a balance between risk and cost. It depends on many considerations, including system complexity, organizational complexity, reporting frequency, how many contractors, program office size and make up, contractor past performance, political visibility, and contract type.

14.4 SUMMARY POINTS

- Management of technical activities requires use of three basic types of metrics: product metrics that track the development of the product, earned value which tracks conformance to the planned schedule and cost, and management process metrics that track management activities.

- Measurement, evaluation and control of metrics is accomplished through a system of periodic reporting that must be planned, established, and monitored to assure metrics are measured properly, evaluated, and the resulting data disseminated.

- TPMs are performance based product metrics that track progress through measurement of key technical parameters. They are important to the systems engineering process because they connect operational requirements to measurable design characteristics and help assess how well the effort is meeting those requirements. TPMs are required for all programs covered by DoD 5000.2-R.

SUPPLEMENT 14-A

TECHNICAL PERFORMANCE MEASUREMENT

Technical Performance Measurement (TPM) is an analysis and control technique that is used to: (1) project the probable performance of a selected technical parameter over a period of time, (2) record the actual performance observed of the selected parameter, and (3) through comparison of actual versus projected performance, assist the manager in decision making. A well thought out program of technical performance measures provides an early warning of technical problems and supports assessments of the extent to which operational requirements will be met, as well as assessments of the impacts of proposed changes in system performance.

TPMs generally take the form of both graphic displays and narrative explanations. The graphic, an example of which is shown in Figure 14-2, shows the projected behavior of the selected parameter as a function of time, and further shows actual observations, so that deviations from the planned profile can be assessed. The narrative portion of the report should explain the graphic, addressing the reasons for deviations from the planned profile, assessing the seriousness of those deviations, explaining actions underway to correct the situation if required, and projecting future performance, given the current situation.

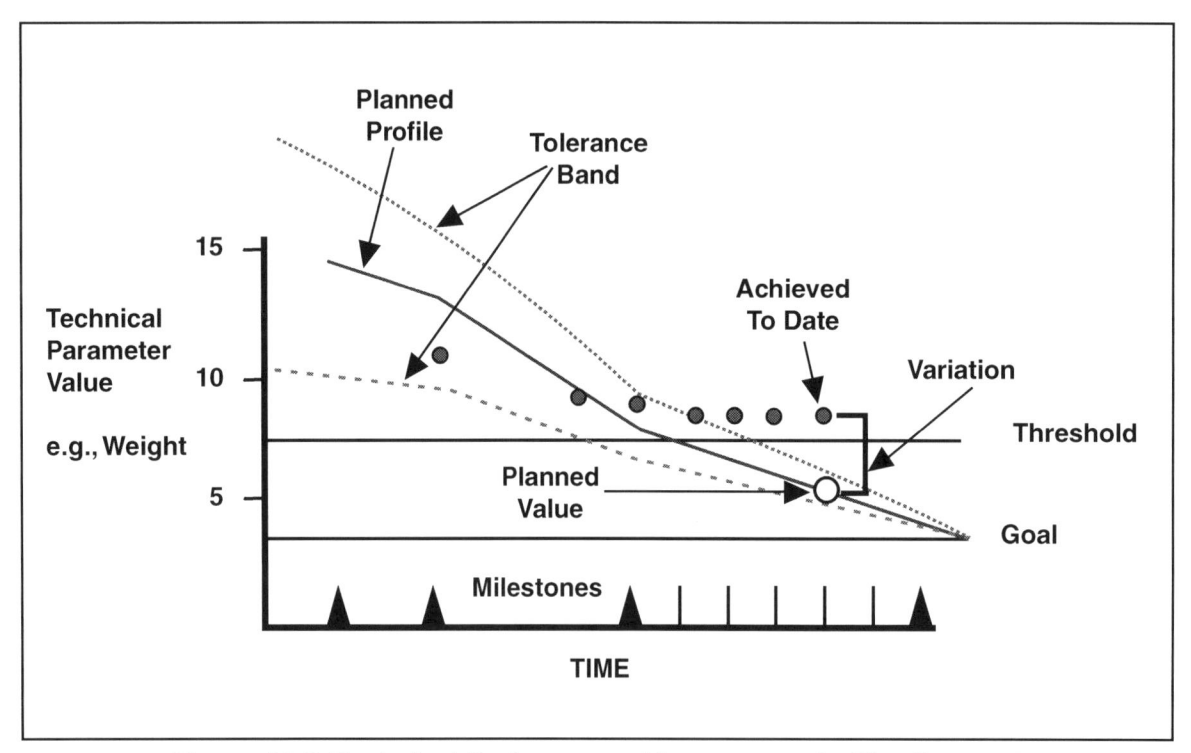

Figure 14-2. Technical Performance Measurement – The Concept

Parameters to be tracked are typically based on the combined needs of the government and the contractor. The government program office will need a set of TPMs which provide visibility into the technical performance of key elements of the WBS, especially those which are cost drivers on the program, lie on the critical path, or which represent high risk items.

The TPMs selected for delivery to the government are expected to be traceable to the needs of the operational user. The contractor will generally track more items than are reported to the government, as the contractor needs information at a more detailed level than does the government program office.

TPM reporting to the government is a contractual issue, and those TPMs on which the government receives reports are defined as contract deliverables in the contract data requirements list. Which parameters are selected for reporting depends on a number of issues, among which are resources to purchase TPMs, the availability of people to review and follow the items, the complexity of the system involved, the phase of development, and the contractor's past experience with similar systems.

A typical TPM graphic will take a form somewhat like that previously shown. The actual form of the projected performance profile and whether or not tolerance bands are employed will be a function of the parameter selected and the needs of the program office.

Another important consideration is the relationship between the TPM program and risk management. Generally, the parameters selected for tracking should be related to the risk areas on the program. If a particular element of the design has been identified as a risk area, then parameters should be selected which will enable the manager to track progress in that area. For example, if achieving a required aircraft range is considered to be critical and a risk area, then tracking parameters that provide insight into range would be selected, such as aircraft weight, specific fuel consumption, drag, etc. Furthermore, there should be consistency between TPMs and the Critical Technical Parameters associated with formal testing, although the TPM program will not normally be limited just to those parameters identified as critical for test purposes.

Government review and follow up of TPMs are appropriate on a periodic basis when submitted by the contractor, and at other major technical events such as at technical reviews, test events, and program management reviews.

While TPMs are expected to be traceable to the needs of the user, they must be concrete technical parameters that can be projected and tracked. For example, an operational user may have a requirement for survivability under combat conditions. Survivability is not, in and of itself, a measurable parameter, but there are important technical parameters that determine survivability, such as radar cross section (RCS) and speed. Therefore, the technical manager might select and track RCS and speed as elements for TPM reporting. The decision on selection of parameters for TPM tracking must also take into consideration the extent to which the parameter behavior can be projected (profiled over a time period) and whether or not it can actually be measured. If the parameter cannot be profiled, measured, or is not critical to program success, then the government, in general, should not select it for TPM tracking. The WBS structure makes an excellent starting point for consideration of parameters for TPM tracking (see Figure 14-3).

A substantial effort has taken place in recent years to link TPMs with Earned Value Management in a way that would result in earned value calculations that reflect the risks associated with achieving technical performance. The approach used establishes statistical probability of achieving a projected level of performance on the TPM profile based on a statistical analysis of actual versus planned performance. Further information is available on the Internet at http://www.acq.osd.mil/api/tpm/.

In summary, TPMs are an important tool in the program manager's systems analysis and control toolkit. They provide an early warning about deviations in key technical parameters, which, if not controlled, can impact system success in meeting user needs. TPMs should be an integral part of both

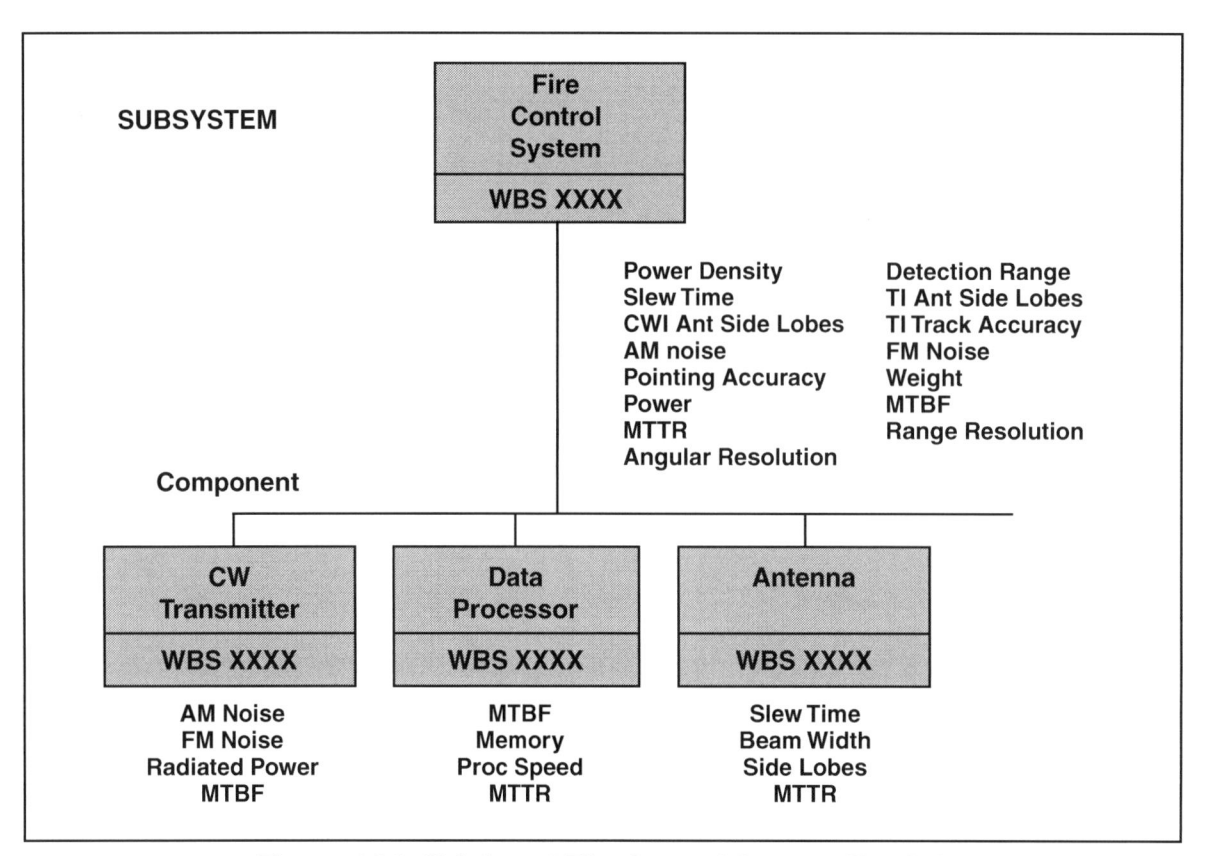

```
                          Fire
      SUBSYSTEM        Control
                        System
                       WBS XXXX

                              Power Density        Detection Range
                              Slew Time            TI Ant Side Lobes
                              CWI Ant Side Lobes   TI Track Accuracy
                              AM noise             FM Noise
                              Pointing Accuracy    Weight
                              Power                MTBF
                              MTTR                 Range Resolution
                              Angular Resolution
      Component

           CW              Data              Antenna
        Transmitter      Processor
        WBS XXXX         WBS XXXX           WBS XXXX

        AM Noise          MTBF              Slew Time
        FM Noise          Memory            Beam Width
        Radiated Power    Proc Speed        Side Lobes
        MTBF              MTTR              MTTR
```

Figure 14-3. Shipboard Fire Control System (Partial)

periodic program reporting and management follow-up, as well as elements for discussion in technical reviews and program management reviews. By thoughtful use of a good program of TPM, the manager, whether technically grounded or not, can make perceptive judgments about system technical performance and can follow up on contractor plans and progress when deviations occur.

Relevant Terms

Achievement to date — Measured or estimated progress plotted and compared with planned progress by designated milestone date.

Current estimate — Expected value of a technical parameter at contract completion.

Planned value — Predicted value of parameter at a given point in time.

Planned profile — Time phased projected planned values.

Tolerance band — Management alert limits representing projected level of estimating error.

Threshold — Limiting acceptable value, usually contractual.

Variance — Difference between the planned value and the achievement-to-date derived from analysis, test, or demonstration.

CHAPTER 15

RISK MANAGEMENT

15.1 RISK AS REALITY

Risk is inherent in all activities. It is a normal condition of existence. Risk is the potential for a negative future reality that may or may not happen. Risk is defined by two characteristics of a possible negative future event: probability of occurrence (whether something will happen), and consequences of occurrence (how catastrophic if it happens). If the probability of occurrence is not known then one has *uncertainty*, and the risk is undefined.

Risk is not a problem. It is an understanding of the level of threat due to *potential* problems. A problem is a consequence that has already occurred.

In fact, knowledge of a risk is an opportunity to avoid a problem. Risk occurs whether there is an attempt to manage it or not. Risk exists whether you acknowledge it, whether you believe it, whether if it is written down, or whether you understand it. Risk does not change because you hope it will, you ignore it, or your boss's expectations do not reflect it. Nor will it change just because it is contrary to policy, procedure, or regulation. Risk is neither good nor bad. It is just how things are. Progress and opportunity are companions of risk. In order to make progress, risks must be understood, managed, and reduced to acceptable levels.

Types of Risk in a Systems Engineering Environment

Systems engineering management related risks could be related to the system products or to the process of developing the system. Figure 15-1 shows the decomposition of system development risks.

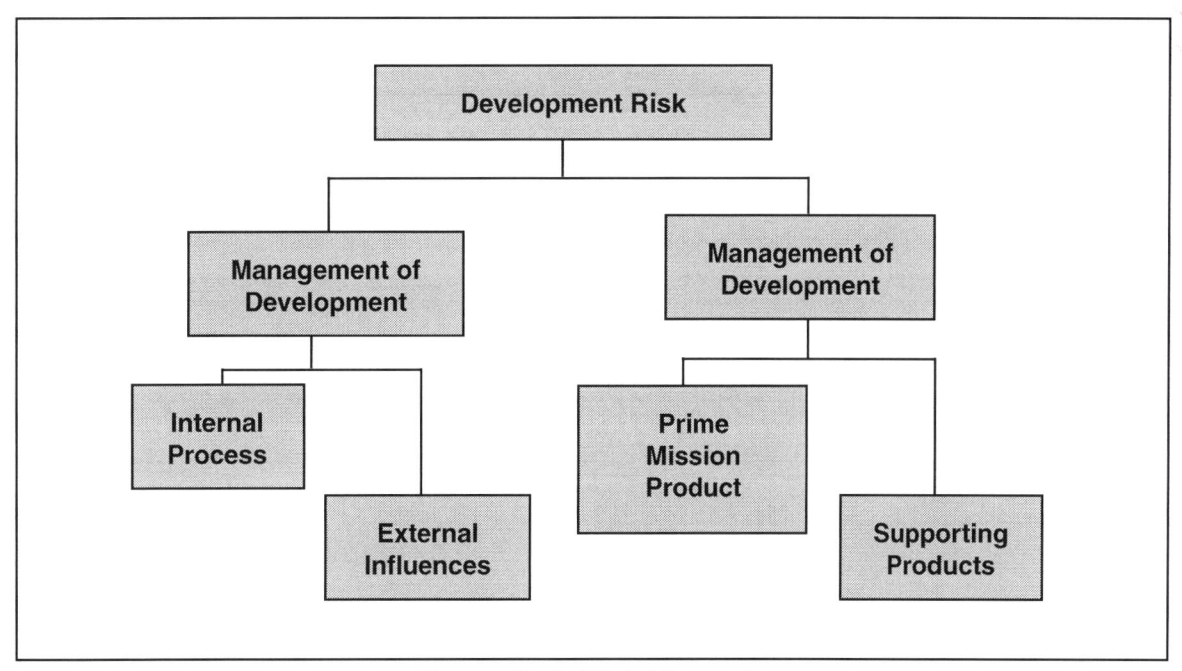

Figure 15-1. Risk Hierarchy

Risks related to the system development generally are traceable to achieving life cycle customer requirements. Product risks include both end product risks that relate to the basic performance and cost of the system, and to enabling products that relate to the products that produce, maintain, support, test, train, and dispose of the system.

Risks relating to the management of the development effort can be technical management risk or risk caused by external influences. Risks dealing with the internal technical management include those associated with schedules, resources, work flow, on time deliverables, availability of appropriate personnel, potential bottlenecks, critical path operations and the like. Risks dealing with external influences include resource availability, higher authority delegation, level of program visibility, regulatory requirements, and the like.

15.2 RISK MANAGEMENT

Risk management is an organized method for identifying and measuring risk and for selecting, developing, and implementing options for the handling of risk. It is a process, not a series of events. Risk management depends on risk management planning, early identification and analysis of risks, continuous risk tracking and reassessment, early implementation of corrective actions, communication, documentation, and coordination. Though there are many ways to structure risk management, this book will structure it as having four parts: Planning, Assessment, Handling, and Monitoring. As depicted in Figure 15-2 all of the parts are interlocked to demonstrate that after initial planning the parts begin to be dependent on each other. Illustrating this, Figure 15-3 shows the key control and feedback relationships in the process.

Risk Planning

Risk Planning is the continuing process of developing an organized, comprehensive approach to risk management. The initial planning includes establishing a strategy; establishing goals and objectives; planning assessment, handling, and monitoring activities; identifying resources, tasks, and responsibilities; organizing and training risk management IPT members; establishing a method to track risk items; and establishing a method to

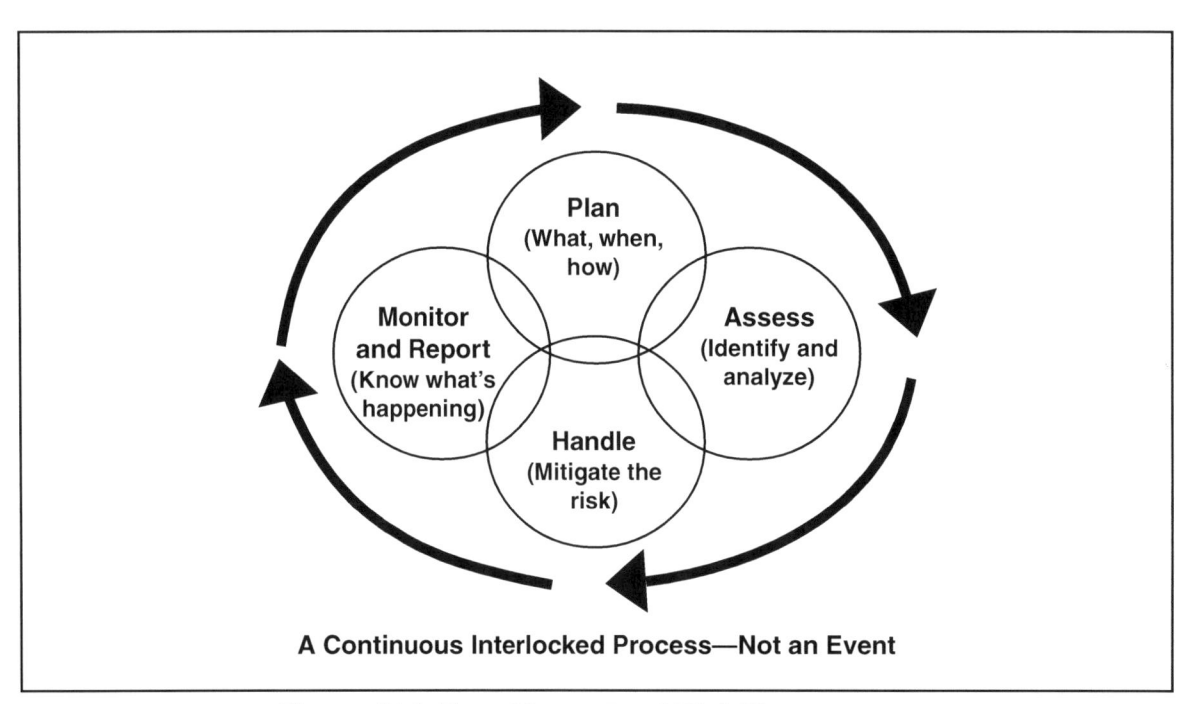

Figure 15-2. Four Elements of Risk Management

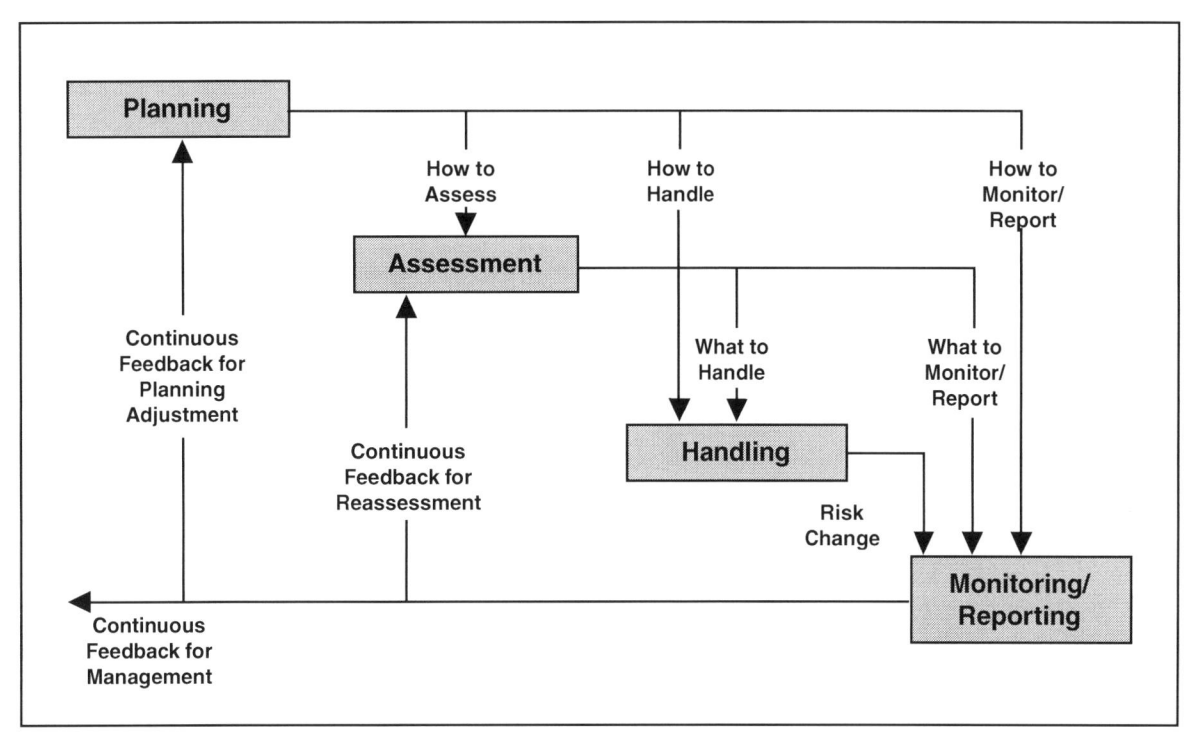

Figure 15-3. Risk Management Control and Feedback

document and disseminate information on a continuous basis.

In a systems engineering environment risk planning should be:

- Inherent (imbedded) in systems engineering planning and other related planning, such as producibility, supportability, and configuration management;

- A documented, continuous effort;

- Integrated among all activities;

- Integrated with other planning, such as systems engineering planning, supportability analysis, production planning, configuration and data management, etc.;

- Integrated with previous and future phases; and

- Selective for each Configuration Baseline.

Risk is altered by time. As we try to control or alter risk, its probability and/or consequence will change. Judgment of the risk impact and the method of handling the risk must be reassessed and potentially altered as events unfold. Since these events are continually changing, the planning process is a continuous one.

Risk Assessment

Risk assessment consists of *identifying* and *analyzing* the risks associated with the life cycle of the system.

Risk Identification Activities

Risk identification activities establish what risks are of concern. These activities include:

- Identifying risk/uncertainty sources and drivers,

- Transforming uncertainty into risk,

- Quantifying risk,

- Establishing probability, and

- Establishing the priority of risk items.

As shown by Figure 15-4 the initial identification process starts with an identification of potential risk items in each of the four risk areas. Risks related to the system performance and supporting products are generally organized by WBS and initially determined by expert assessment of teams and individuals in the development enterprise. These risks tend to be those that require follow-up quantitative assessment. Internal process and external influence risks are also determined by expert assessment within the enterprise, as well as through the use of risk area templates similar to those found in DoD 4245.7-M. The DoD 4245.7-M templates describe the risk areas associated with system acquisition management processes, and provide methods for reducing traditional risks in each area. These templates should be tailored for specific program use based on expert feedback.

After identifying the risk items, the risk level should be established. One common method is through the use of a matrix such as shown in Figure 15-5. Each item is associated with a block in the matrix to establish relative risk among them.

On such a graph risk increases on the diagonal and provides a method for assessing relative risk. Once the relative risk is known, a priority list can be established and risk analysis can begin.

Risk identification efforts can also include activities that help define the probability or consequences of a risk item, such as:

- Testing and analyzing uncertainty away,

- Testing to understand probability and consequences, and

- Activities that quantify risk where the qualitative nature of high, moderate, low estimates are insufficient for adequate understanding.

Risk Analysis Activities

Risk analysis activities continue the assessment process by refining the description of identified risk event through isolation of the cause of risk, determination of the full impact of risk, and the

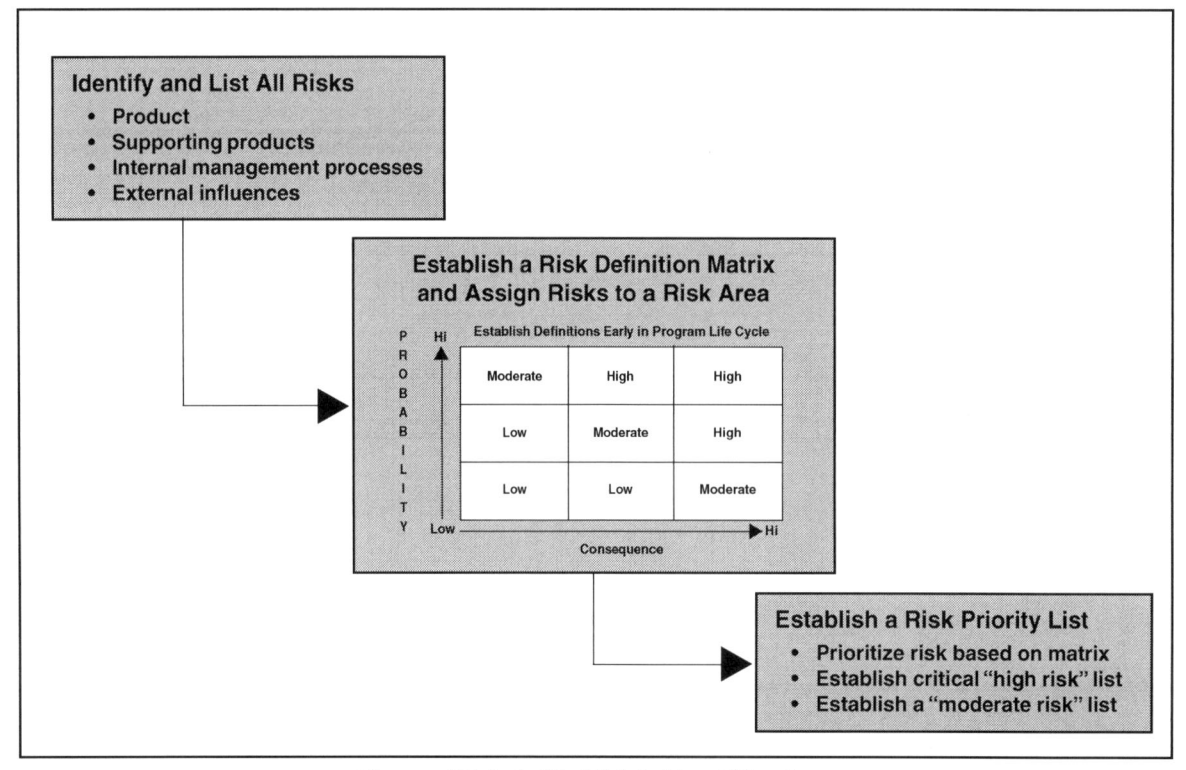

Figure 15-4. Initial Risk Identificaiton

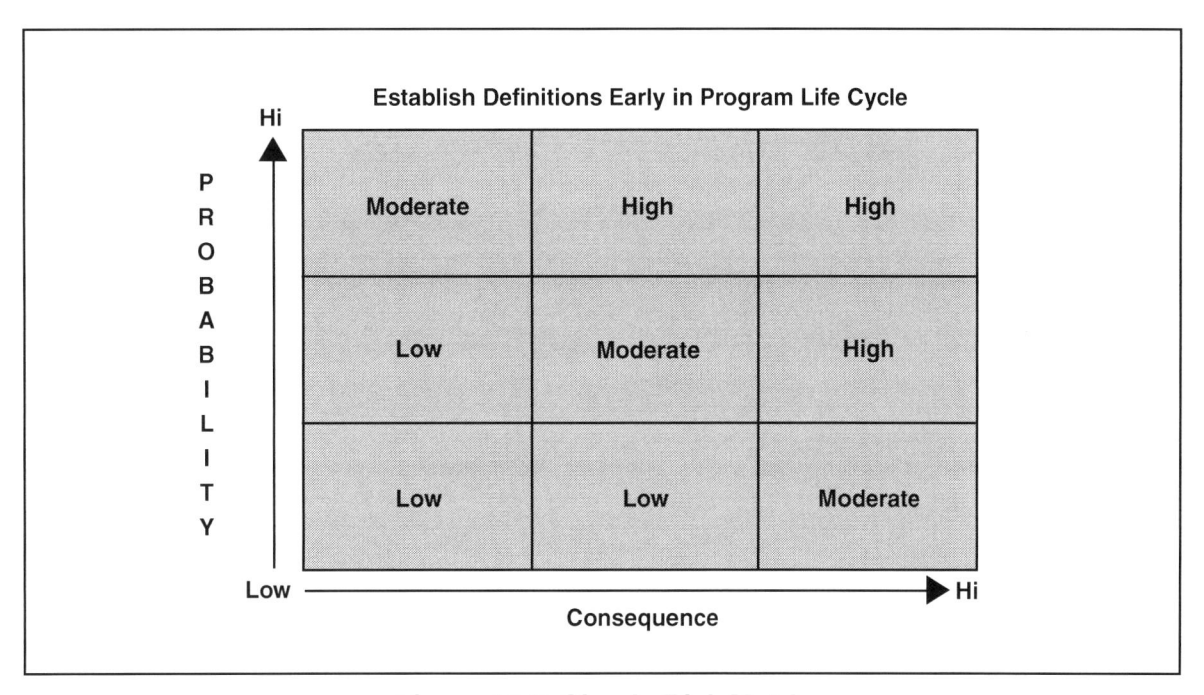

Figure 15-5. Simple Risk Matrix

determination and choose of alternative courses of action. They are used to determine what risk should be tracked, what data is used to track risk, and what methods are used to handle the risk.

Risk analysis explores the options, opportunities, and alternatives associated with the risk. It addresses the questions of how many legitimate ways the risk could be dealt with and the best way to do so. It examines sensitivity, and risk interrelationships by analyzing impacts and sensitivity of related risks and performance variation. It further analyzes the impact of potential and accomplished, external and internal changes.

Risk analysis activities that help define the scope and sensitivity of the risk item include finding answers to the following questions:

- If something changes, will risk change faster, slower, or at the same pace?

- If a given risk item occurs, what collateral effects happen?

- How does it affect other risks?

- How does it affect the overall situation?

- Development of a watch list (prioritized list of risk items that demand constant attention by management) and a set of metrics to determine if risks are steady, increasing, or decreasing.

- Development of a feedback system to track metrics and other risk management data.

- Development of quantified risk assessment.

Quantified risk assessment is a formal quantification of probabilities of occurrence and consequences using a top-down structured process following the WBS. For each element, risks are assessed through analysis, simulation and test to determine statistical probability and specific conditions caused by the occurrence of the consequence.

Cautions in Risk Assessments

Reliance solely on numerical values from simulations and analysis should be avoided. Do not lose sight of the actual source and consequences of the risks. Testing does not eliminate risk. It only provides data to assess and analyze risk. Most of all, beware of manipulating relative numbers, such as 'risk index" or "risk scales," even when based on expert opinion, as quantified data. They are important information, but they are largely subjective and relative; they do not necessarily define risk accurately. Numbers such as these should always be the subject of a sensitivity analysis.

Risk Handling

Once the risks have been categorized and analyzed, the process of handling those risks is initiated. The prime purpose of risk handling activities is to mitigate risk. Methods for doing this are numerous, but all fall into four basic categories:

- Risk Avoidance,

- Risk Control,

- Risk Assumption, and

- Risk Transfer.

Avoidance
To avoid risk, remove requirements that represent uncertainty and high risk (probability or consequence.) Avoidance includes trading off risk for performance or other capability, and it is a key activity during requirements analysis. Avoidance requires understanding of priorities in requirements and constraints. Are they mission critical, mission enhancing, nice to have, or "bells and whistles?"

Control
Control is the deliberate use of the design process to lower the risk to acceptable levels. It requires the disciplined application of the systems engineering process and detailed knowledge of the technical area associated with the design. Control techniques are plentiful and include:

- Multiple concurrent design to provide more than one design path to a solution,

- Alternative low-risk design to minimize the risk of a design solution by using the lowest-risk design option,

- Incremental development, such as preplanned product improvement, to dissociate the design from high-risk components that can be developed separately,

- Technology maturation that allows high-risk components to be developed separately while the basic development uses a less risky and lower-performance temporary substitute,

- Test, analyze and fix that allows understanding to lead to lower risk design changes. (Test can be replaced by demonstration, inspection, early prototyping, reviews, metric tracking, experimentation, models and mock-ups, simulation, or any other input or set of inputs that gives a better understanding of the risk),

- Robust design that produces a design with substantial margin such that risk is reduced, and

- The open system approach that emphasizes use of generally accepted interface standards that provide proven solutions to component design problems.

Acceptance
Acceptance is the deliberate acceptance of the risk because it is low enough in probability and/or consequence to be reasonably assumed without impacting the development effort. Key techniques for handling accepted risk are budget and schedule reserves for unplanned activities and continuous assessment (to assure accepted risks are maintained at acceptance level). The basic objective of risk management in systems engineering is to reduce all risk to an acceptable level.

The strong budgetary strain and tight schedules on DoD programs tends to reduce the program manager's and system engineer's capability to provide reserve. By identifying a risk as acceptable, the worst-case outcome is being declared acceptable. Accordingly, the level of risk considered acceptable should be chosen very carefully in a DoD acquisition program.

Transfer

Transfer can be used to reduce risk by moving the risk from one area of design to another where a design solution is less risky. Examples of this include:

- Assignment to hardware (versus software) or vice versa; and

- Use of functional partitioning to allocate performance based on risk factors.

Transfer is most associated with the act of assigning, delegating, or paying someone to assume the risk. To some extent transfer always occurs when contracting or tasking another activity. The contract or tasking document sets up agreements that can transfer risk from the government to contractor, program office to agency, and vice versa. Typical methods include insurance, warranties, and incentive clauses. Risk is never truly transferred. If the risk isn't mitigated by the delegated activity it still affects your project or program.

Key areas to review before using transfer are:

- How well can the delegated activity handle the risk? Transfer is effective only to the level the risk taker can handle it.

- How well will the delegated activity solution integrate into your project or program? Transfer is effective only if the method is integrated with the overall effort. For example, is the warranty action coordinated with operators and maintainers?

- Was the method of tasking the delegated activity proper? Transfer is effective only if the transfer mechanism is valid. For example, can incentives be "gamed?"

- Who has the most control over the risk? If the project or program has no or little control over the risk item, then transfer should be considered to delegate the risk to those most likely to be able to control it.

Monitoring and Reporting

Risk monitoring is the continuous process of tracking and evaluating the risk management process by metric reporting, enterprise feedback on watch list items, and regular enterprise input on potential developing risks. (The metrics, watch lists, and feedback system are developed and maintained as an assessment activity.) The output of this process is then distributed throughout the enterprise, so that all those involved with the program are aware of the risks that affect their efforts and the system development as a whole.

Special Case – Integration as Risk

Integration of technologies in a complex system is a technology in itself! Technology integration during design may be a high-risk item. It is not normally assessed or analyzed as a separately identified risk item. If integration risks are not properly identified during development of the functional baseline, they will demonstrate themselves as serious problems in the development of the product baseline.

Special Case – Software Risk

Based on past history, software development is often a high-risk area. Among the causes of performance, schedule, and cost deficiencies have been:

- Imperfect understanding of operational requirements and its translation into source instructions,

- Risk tracking and handling,

- Insufficient comprehension of interface constraints, and

- Lack of sufficient qualified personnel.

Risk Awareness

All members of the enterprise developing the system must understand the need to pay attention to the existence and changing nature of risk.

Consequences that are unanticipated can seriously disrupt a development effort. The uneasy feeling that something is wrong, despite assurances that all is fine may be valid. These kinds of intuitions have allowed humanity to survive the slings and arrows of outrageous fortune throughout history. Though generally viewed as non-analytical, these apprehensions should not be ignored. Experience indicates those non-specific warnings have validity, and should be quantified as soon as possible.

15.3 SUMMARY POINTS

- Risk is inherent in all activities.

- Risk is composed of knowledge of two characteristics of a possible negative future event: probability of occurrence and consequences of occurrence.

- Risk management is associated with a clear understanding of probability.

- Risk management is an essential and integral part of technical program management (systems engineering).

- Risks and uncertainties must be identified, analyzed, handled, and tracked.

- There are four basic ways of handling risk: avoidance, transfer, acceptance, and control.

- Program risks are classified as low, moderate, or high depending on consequences and probability of occurrence. Risk classification should be based on quantified data to the extent possible.

SUPPLEMENT 15-A

RISK MANAGEMENT IN DOD ACQUISITION

Policy

DoD policy is quite clear in regard to risk management: it must be done.

The PM shall identify the risk areas in the program and integrate risk management within overall program management. (DoD 5000.2-R.)

In addition, DoDD 5000.4 identifies risk and cost analysis as a responsibility of the program manager.

Risk Management View

A DSMC study indicates that major programs which declared moderate risk at Milestone B have been more successful in terms of meeting cost and schedule goals than those which declared low risk (DSMC TR 2-95). This strongly implies that program offices that understand and respect risk management will be more successful. For this reason, the program office needs to adopt a systems-level view of risk. The systems engineer provides this view. Systems Engineering is the cornerstone of program office risk management program because it is the connection to realistic assessment of product maturity and development, and the product is, in the final analysis, what system acquisition is really about.

However, the program office has external risks to deal with as well as the internal risks prevalent in the development process. The Systems Engineer has to provide the program manager internal risk data in a manner that aids the handling of the external risks. In short, the systems engineer must present bad news such that it is reasonable and compelling to higher levels of authority. See Chapter 20 for further discussion on this topic.

Factoring Risk Management into the Process

Risk management, as an integral part of the overall program planning and management process, is enhanced by applying a controlled, consistent, approach to systems engineering and using integrated teams for both product development and management control. Programs should be transitioned to the next phase only if risk is at the appropriate level. Know the risk drivers behind the estimates. By its nature there are always subjective aspects to assessing and analyzing risk at the system level, even though they tend to be represented as quantitative and/or analytically objective.

Risk and Phases

Risk management begins in the Concept and Technology Development phase. During Concept Exploration initial system level risk assessments are made. Unknown-unknowns, uncertainty, and some high-risk elements are normal and expected. When substantial technical risk exists, the Component Advanced Development stage is appropriate, and is included in the life-cycle process specifically as an opportunity to address and reduce risks to a level that are consistent with movement into systems acquisition.

The S&T community has a number of vehicles available that are appropriate for examining technology in application and for undertaking risk reduction activities. These include Advanced Technology Demonstrations, Advanced Concept Technology Demonstrations, as well as Joint Warfighting Experiments. The focus of the activities undertaken during these risk reduction stages include:

- Testing, analyzing, or mitigating system and subsystem uncertainty and high risk out of the program.

- Demonstrating technology sufficient to uncover system and subsystem unknown-unknowns (especially for integration).

- Planning for risk management during the transition to and continuation of systems acquisition during the System Development and Demonstration phase, especially handling and tracking of moderate risk.

System Development and Demonstration requires the application of product and manufacturing engineering, which can be disrupted if the technology development is not sufficient to support engineering development. Risk management in during this phase emphasizes:

- Reduction and control of moderate risks,

- All risks under management including emerging ones, and

- Maintenance of risk levels and reaction to problems.

Objective Assessment of Technology

The revised acquisition process has been deliberately structured to encourage and allow programs to progress through appropriate risk reduction stages and phases, based on an objective assessment of the maturity levels associated with the products and systems under development. It is therefore, particularly important that program managers and their staffs ensure that the decisions made regarding recommendations to proceed, and the paths to be taken, be based on as impartial and objective opinions as possible. The temptation is always to move ahead and not to delay to improve the robustness of a given product or system. When systems are hurried into engineering development and production, in spite of the fact that the underlying technologies require further development, history indicates that the results will eventually show the fallacy of speed over common sense. And to fix the problem in later stages of development—or even after deployment—can be hugely expensive in terms of both monetary cost and human lives.

The prevailing presumption at Milestone B is that the system is ready for engineering development. After this, the acquisition community generally assumes that risk is moderate to low, that the technology is "available." There is evidence to support the assertion that programs often progress into engineering development with risks that actually require substantial exploratory and applied research and development to bring them to the moderate levels of risk or lower. One approach that has proven successful in making objective risk assessments is the use of independent evaluation teams. Groups that have no pre-determined interest to protect or axe to grind are often capable of providing excellent advice regarding the extent to which a system is ready to proceed to the next level of development and subsequent phases.

Risk Classification on the System (Program) Level

Classification definitions should be established early and remain consistent throughout the program. The program office should assess the risks of achieving performance, schedule, and cost in clear and accurate terms of both probability and consequence. Where there is disagreement about the risk, assessment efforts should be immediately increased. Confusion over risk is the worst program risk, because it puts in doubt the validity of the risk management process, and therefore, whether program reality is truly understood.

The system level risk assessment requires integration and interpretation of the quantified risk assessment of the parts. This requires reasonable judgement. Because integration increases the potential for risk, it is reasonable to assume overall risk is not better than the sum of objective data for the parts.

Reality Versus Expectations

Program managers are burdened with the expectations of superiors and others that have control over the program office's environment. Pressure to accommodate these expectations is high. If the systems engineer cannot communicate the reality of risk in terms that are understandable, acceptable, or sufficiently verifiable to management, then these pressures may override vertical communication of actual risk.

Formal systems engineering with risk management incorporated can provide the verifiable information. However, the systems engineer also has the responsibility to adequately explain probability and consequences such that the program manager can accept the reality of the risk and override higher level expectations.

Uncertainty is a special case, and very dangerous in an atmosphere of high level expectations. Presentation of uncertainty issues should strongly emphasize consequences, show probability trends, and develop "most likely" alternatives for probability.

SUPPLEMENT 15-B

MODEL FOR SYSTEM LEVEL RISK ASSESSMENT

The following may be used to assist in making preliminary judgments regarding risk classifications:

	Low Risk	Moderate Risk	High Risk
Consequences	Insignificant cost, schedule, or technical impact	Affects program objectives, cost, or schedule; however cost, schedule, performance are achievable	Significant impact, requiring reserve or alternate courses of action to recover
Probability of Occurrence	Little or no estimated likelihood	Probability sufficiently high to be of concern to management	High likelihood of occurrence
Extent of Demonstration	Full-scale, integrated technology has been demonstrated previously	Has been demonstrated but design changes, tests in relevant environments required	Significant design changes required in order to achieve required/desired results
Existence of Capability	Capability exists in known products; requires integration into new system	Capability exists, but not at performance levels required for new system	Capability does not currently exist

Also see Technology Readiness Levels matrix in Chapter 2

PART 4

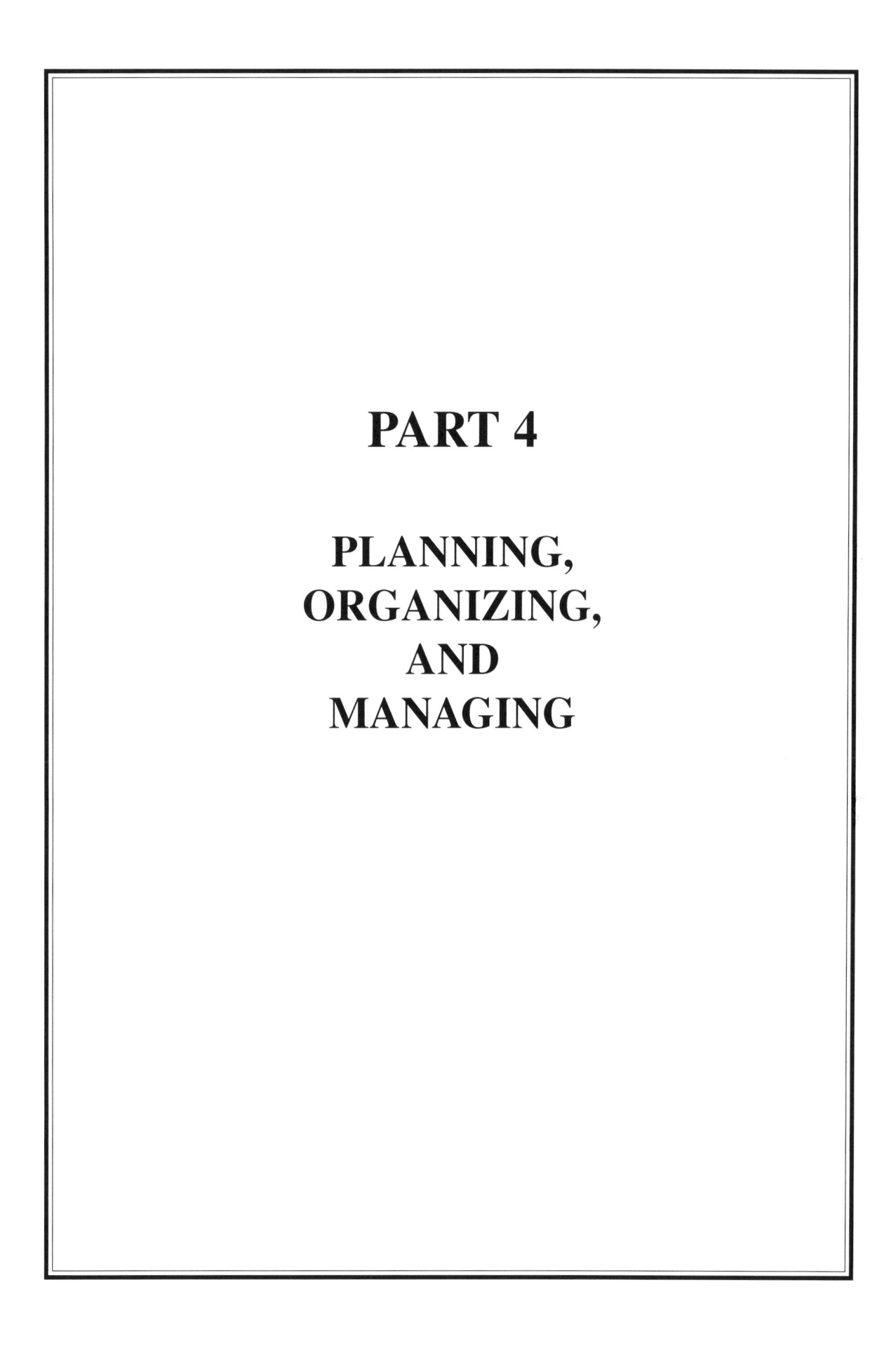

PLANNING, ORGANIZING, AND MANAGING

CHAPTER 16

SYSTEMS ENGINEERING PLANNING

16.1 WHY ENGINEERING PLANS?

Systems engineering planning is an activity that has direct impact on acquisition planning decisions and establishes the feasible methods to achieve the acquisition objectives. Management uses it to:

- Assure that all technical activities are identified and managed,

- Communicate the technical approach to the broad development team,

- Document decisions and technical implementation, and

- Establish the criteria to judge how well the system development effort is meeting customer and management needs.

Systems engineering planning addresses the scope of the technical effort required to develop the system. The basic questions of "who will do what" and "when" are addressed. As a minimum, a technical plan describes what must be accomplished, how systems engineering will be done, how the effort will be scheduled, what resources are needed, and how the systems engineering effort will be monitored and controlled. The planning effort results in a management-oriented document covering the implementation of program requirements for system engineering, including technical management approaches for subsequent phases of the life cycle. In DoD it is an exercise done on a systems level by the government, and on a more detailed level by contractors.

Technical/Systems Engineering Planning

Technical planning may be documented in a separate engineering management plan or incorporated into a broad, integrated program management plan. This plan is first drafted at project or program inception during the early requirements analysis effort. Requirements analysis and technical planning are inherently linked, because requirements analysis establishes an understanding of what must be provided. This understanding is fundamental to the development of detailed plans.

To be of utility, systems engineering plans must be regularly updated. To support management decision making, major updates will usually occur at least just before major management milestone decisions. However, updates must be performed as necessary between management milestones to keep the plan sufficiently current to achieve its purpose of information, communication, and documentation.

16.2 ELEMENTS OF TECHNICAL PLANS

Technical plans should include sufficient information to document the purpose and method of the systems engineering effort. Plans should include the following:

- An introduction that states the purpose of the engineering effort and a description of the system being developed,

- A technical strategy description that ties the engineering effort to the higher-level management planning,

- A description of how the systems engineering process will be tailored and structured to complete the objectives stated in the strategy,

- An organization plan that describes the organizational structure that will achieve the engineering objectives, and

- A resource plan that identifies the estimated funding and schedule necessary to achieve the strategy.

Introduction

The introduction should include:

Scope: The scope of the plan should provide information concerning what part of the big picture the plan covers. For example, if the plan were a DoD program office plan, it would emphasize control of the higher-level requirements, the system definition (functional baseline), and all activities necessary for system development. On the other hand, a contractor's plan would emphasize control of lower-level requirements, preliminary and detail designs (allocated and product baselines), and activities required and limited by the contractual agreement.

Description: The description of the system should:

- Be limited to an executive summary describing those features that make the system unique,

- Include a general discussion of the system's operational functions, and

- Answer the question "What is it and what will it do?"

Focus: A guiding focus for the effort should be provided to clarify the management vision for the development approach. For example, the focus may be *lowest cost to obtain threshold requirements, superior performance within budget, superior standardization for reduced logistics, maximum use of the open systems approach to reduce cost*, or the like. A focus statement should:

- Be a single objective to avoid confusion,

- Be stated simply to avoid misinterpretation, and

- Have high-level support.

Purpose: The purpose of the engineering effort should be described in general terms of the outputs, both end products and life-cycle enabling products that are required. The stated purpose should answer the question, "What does the engineering effort have to produce?"

Technical Strategy

The basic purpose of a technical strategy is to link the development process with the acquisition or contract management process. It should include:

- Development phasing and associated baselining,

- Key engineering milestones to support risk management and business management milestones,

- Associated parallel developments or product improvement considerations, and

- Other management generated constraints or high-visibility activities that could affect the engineering development.

Phasing and Milestones: The development phasing and baseline section should describe the approach to phasing the engineering effort, including tailoring of the basic process described in this book and a rationale for the tailoring. The key milestones should be in general keeping with the technical review process, but tailored as appropriate to support business management milestones and the project/program's development phasing. Strategy considerations should also include discussion of how design and verification will phase into production and fielding. This area should identify how production will be phased-in (including use of limited-rate initial production and long lead-time purchases), and that initial support considerations require significant coordination between the user and acquisition community.

Parallel Developments and Product Improvement: Parallel development programs necessary for the system to achieve its objectives should be identified and the relationship between the efforts explained. Any product improvement strategies should also be identified. Considerations such as evolutionary development and preplanned product improvement should be described in sufficient detail to show how they would phase into the overall effort.

Impacts on Strategy

All conditions or constraints that impact the strategy should be identified and the impact assessed. Key points to consider are:

- Critical technologies development,

- Cost As an Independent Variable (CAIV), and

- Any business management directed constraint or activity that will have a significant influence on the strategy.

Critical Technologies: Discussion of critical technology should include:

- Risk associated with critical technology development and its impact on the strategy,

- Relationship to baseline development, and

- Potential impact on the overall development effort.

Cost As an Independent Variable: Strategy considerations should include discussion of how CAIV will be implemented, and how it will impact the strategy. It should discuss how unit cost, development cost, life cycle cost, total ownership cost, and their interrelationships apply to the system development. This area should focus on how these costs will be balanced, how they will be controlled, and what impact they have on the strategy and design approach.

Management Issues: Management issues that pose special concerns for the development strategy could cover a wide range of possible issues. In general, management issues identified as engineering strategy issues are those that impact the ability to support the management strategy. Examples would include:

- Need to combine developmental phases to accommodate management driven schedule or resource limitations,

- Risk associated with a tight schedule or limited budget,

- Contractual approach that increases technical risk, and

- Others of a similar nature.

Management-dictated technical activities—such as use of M&S, open systems, IPPD, and others—should not be included as a strategy issue unless they impact the overall systems engineering strategy to meet management expectations. The strategy discussion should lay out the plan, how it dovetails with the management strategy, and how management directives impact it.

Systems Engineering Processes

This area of the planning should focus on how the system engineering processes will be designed to support the strategy. It should include:

- Specific methods and techniques used to perform the steps and loops of the systems engineering process,

- Specific system analysis and control tools and how they will be used to support step and loop activities, and

- Special design considerations that must be integrated into the engineering effort.

Steps and Loops: The discussion of how the systems engineering process will be done should show the specific procedures and products that will ensure:

- Requirements are understood prior to the flow-down and allocation of requirements,

- Functional descriptions are established before designs are formulated,

- Designs are formulated that are traceable to requirements,

- Methods exist to reconsider previous steps, and

- Verification processes are in place to ensure that design solutions meet needs and requirements.

This planning area should address each step and loop for each development phase, include identification of the step-specific tools (Functional Flow Block Diagrams, Timeline Analysis, etc.) that will be used, and establish the verification approach. The verification discussion should identify all verification activities, the relationship to formal developmental T&E activities, and independent testing activities (such as operational testing).

Norms of the particular technical area and the engineering processes of the command, agency, or company doing the tasks will greatly influence this area of planning. However, whatever procedures, techniques, and analysis products or models used, they should be compatible with the basic principles of systems engineering management as described earlier in this book.

An example of the type of issue this area would address is the requirements analysis during the system definition phase. Requirements analysis is more critical and a more central focus during system definition than in later phases. The establishment of the correct set of customer requirements at the beginning of the development effort is essential to proper development. Accordingly, the system definition phase requirements analysis demands tight control and an early review to verify the requirements are established well enough to begin the design effort. This process of control and verification necessary for the system definition phase should be specifically described as part of the overall requirements analysis process and procedures.

Analysis and Control: Planning should identify those analysis tools that will be used to evaluate alternative approaches, analyze or assess effectiveness, and provide a rigorous quantitative basis for selecting performance, functional, and design requirements. These processes can include trade studies, market surveys, M&S, effectiveness analyses, design analyses, QFD, design of experiments, and others.

Planning must identify the method by which control and feedback will be established and maintained. The key to control is performance-based measurement guided by an event-based schedule. Entrance and exit criteria for the event-driven milestones should be established sufficient to demonstrate proper development progress has been completed. Event-based schedules and exit criteria are further discussed later in this chapter. Methods to maintain feedback and control are developed to monitor progress toward meeting the exit criteria. Common methods were discussed earlier in this book in the chapters on metrics, risk management, configuration management, and technical reviews.

Design Considerations: In every system development there are usually technical activities that require special attention. These may come from management concerns, legal or regulatory directives, social issues, or organizational initiatives. For example, a DoD program office will have to conform to DoDD 5000.2-R, which lists several technical activities that must be incorporated into the development effort. DoD plans should specifically address each issue presented in the Program Design section of DoD 5000.2-R.

In the case of a contractor there may be issues delineated in the contract, promised in the proposal, or established by management that the technical effort must address. The system engineering planning must describe how each of these issues will be integrated into the development effort.

Organization

Systems engineering management planning should identify the basic structure that will develop the system. Organizational planning should address how the integration of the different technical disciplines, primary function managers, and other stakeholders will be achieved to develop the system. This planning area should describe how multidisciplinary teaming would be implemented, that is, how the teams will be organized, tasked, and trained. A systems-level team should be established early to support this effort. Roles, authority, and basic responsibilities of the system-level design team should be specifically described. Establishing the design organization should be one of the initial tasks of the system-level design team. Their basic approach to organizing the effort should be described in the plan. Further information on organizing is contained in a later chapter.

Resources

The plan should identify the budget for the technical development. The funds required should be matrixed against a calendar schedule based on the event-based schedule and the strategy. This should establish the basic development timeline with an associated high-level estimated spending profile. Shortfalls in funding or schedule should be addressed and resolved by increasing funds, extending schedule, or reducing requirements prior to the plan preparation. Remember that future analysis of development progress by management will tend to be based on this budget "promised" at plan inception.

16.3 INTEGRATION OF PLANS – PROGRAM PLAN INTERFACES

Systems engineering management planning must be coordinated with interfacing activities such as these:

- Acquisition Strategy assures that technical plans take into account decisions reflected in the Acquisition Strategy. Conflicts must be identified early and resolved.

- Financial plan assures resources match the needs in the tech plan. Conflicts should be identified early and resolved.

- Test and Evaluation Master Plan (TEMP) assures it complements the verification approach. It should provide an integrated approach to verify that the design configuration will meet customer requirements. This approach should be compatible with the verification approach delineated in the systems engineering plan.

- Configuration management plan assures that the development process will maintain the system baselines and control changes to them.

- Design plans (e.g., electrical, mechanical, structural, etc.) coordinates identification of IPT team composition.

- Integrated logistics support planning and support analysis coordinates total system support.

- Production/Manufacturing plan to coordinate activities concerning design producibility, and follow-on production,

- Quality management planning assures that quality engineering activities and quality management functions are included in system engineering planning,

- Risk management planning establishes and coordinates technical risk management to support total program risk management.

- Interoperability planning assures interoperability suitability issues are coordinated with system engineering planning. (Where interoperability is an especially critical requirement such as, communication or information systems, it should be addressed as a separate issue with separate integrated teams, monitoring, and controls).

- Others such as M&S plan, software development plan, human integration plan, environment, safety and health planning, also interface.

Things to Watch

A well developed technical management plan will include:

- The expected benefit to the user,

- How a total systems development will be achieved using a systems engineering approach,

- How the technical plan complements and supports the acquisition or management business plan,

- How incremental reviews will assure that the development stays on track,

- How costs will be reduced and controlled,

- What technical activities are required and who will perform them,

- How the technical activities relate to work accomplishment and calendar dates,

- How system configuration and risk will be controlled,

- How system integration will be achieved,

- How the concerns of the eight primary life cycle functions will be satisfied,

- How regulatory and contractual requirements will be achieved, and

- The feasibility of the plan, i.e., is the plan practical and executable from a technical, schedule, and cost perspective.

16.4 SUMMARY POINTS

- Systems engineering planning should establish the organizational structure that will achieve the engineering objectives.

- Planning must include event-based scheduling and establish feedback and control methods.

- It should result in important planning and control documents for carrying out the engineering effort.

- It should identify the estimated funding and detail schedule necessary to achieve the strategy.

- Systems engineering planning should establish the proper relationship between the acquisition and technical processes.

APPENDIX 16-A

SCHEDULES

The event-based schedule, sometimes referred to as the Systems Engineering Master Schedule (SEMS) or Integrated Master Schedule (IMS) is a technical event-driven (not time-driven) plan primarily concerned with product and process development. It forms the basis for schedule control and progress measurement, and relates engineering management events and accomplishments to the WBS. These events are identified either in the format of entry and exit events (e.g. initiate PDR, complete PDR) or by using entry and exit criteria for each event. Example exit criteria shown in Figures 16-1 and 16-2.

The program office develops an event-based schedule that represents the overall development effort. This schedule is usually high-level and focused on the completion of events that support the acquisition milestone decision process. An event-based schedule is developed by the contractor to include significant accomplishments that must be completed in order to meet the progress required prior to contract established events. The contractor also includes events, accomplishments, and associated success criteria specifically identified by the contract. DoD program offices can use the contractor's event-based schedule and the

System Requirements Review (SRR)	System Functional Review/Software Spec Review (SFR/SSR)	Preliminary Design Review (PDR)
• Mission Analysis completed • Support Strategy defined • System options decisions completed • Design usage defined • Operational performance requirement defined • Manpower sensitivities completed • Operational architecture available and reviewed	• Installed environments defined • Maintenance concept defined • Preliminary design criteria established • Preliminary design margins established • Interfaces defined/preliminary interface specs completed • Software and software support requirements completed • Baseline support/resources requirements defined • Support equipment capability defined • Technical architecture prepared • System defined and requirements shown to be achievable	• Design analyses/definition completed • Material/parts characterization completed • Design maintainability analysis completed/support requirements defined • Preliminary production plan completed • Make/buy decisions finalized • Breadboard investigations completed • Coupon testing completed • Design margins completed • Preliminary FMECA completed • Software functions and architecture and support defined • Maintenance tasks trade studies completed • Support equipment development specs completed

Figure 16-1. Sample Event-Based Schedule Exit Criteria

Critical Design Review Test Readiness Review (CDR/TRR)	System Verfication Review/ Functional Configuration Audit (SVR/FCA)	Physical Configuration Audit (PCA)
• Parts, materials, processes selected • Development tests completed • Inspection points/criteria completed • Component level FMECA completed • Repair level analysis completed • Facility requirements defined • Software test descriptions completed • Hardware and software hazard analysis completed • Firmware spt completed • Software programmers manual completed • Durability test completed • Maintinability analyses completed • Qualification test procedures approved • Producibility analyses completed	• All verification tasks completed • Durability tests completed • Long lead time items identified • PME and operational training completed • Tech manuals completed • Flight test plan approved • Support and training equipment developed • Fielding analysis completed • Provisioning data verified	• Qualification testing completed • All QA provisions finalized • All manufacturing process requirements and documentation finalized • Product fabrication specifications finalized • Support and training equipment qualification completed • All acceptance test requirements completed • Life management plan completed • System support capability demonstrated • Post production support analysis completed • Final software description document and all user manuals complete

Figure 16-2. Sample Event-Driven Schedule Exit Criteria (continued)

contractor's conformance to it for several purposes: source selection, monitoring contractor progress, technical and other reviews, readiness for option award, incentives/awards determination, progress payments decision, and similar activities.

The event-based schedule establishes the key parameters for determining the progress of a development program. To some extent it controls and interfaces with systems engineering management planning, integrated master schedules and integrated master plans, as well as risk management planning, system test planning, and other key plans which govern the details of program management.

The calendar or detail schedule is a time-based schedule that shows how work efforts will support tasks and events identified in the event-based schedule. It aligns the tasks and calendar dates to show when each significant accomplishment must be achieved. It is a key component for developing Earned Value metrics. The calendar schedule is commonly referred to as the detail schedule, systems engineering detail schedule, or SEDS. The contractor is usually required to maintain the relationship between the event and calendar schedules for contract required activities. Figure 16-3 shows the relationship between the system requirements, the WBS, the contractual requirements, the event-based schedule, and the detail schedule.

Figure 16-3

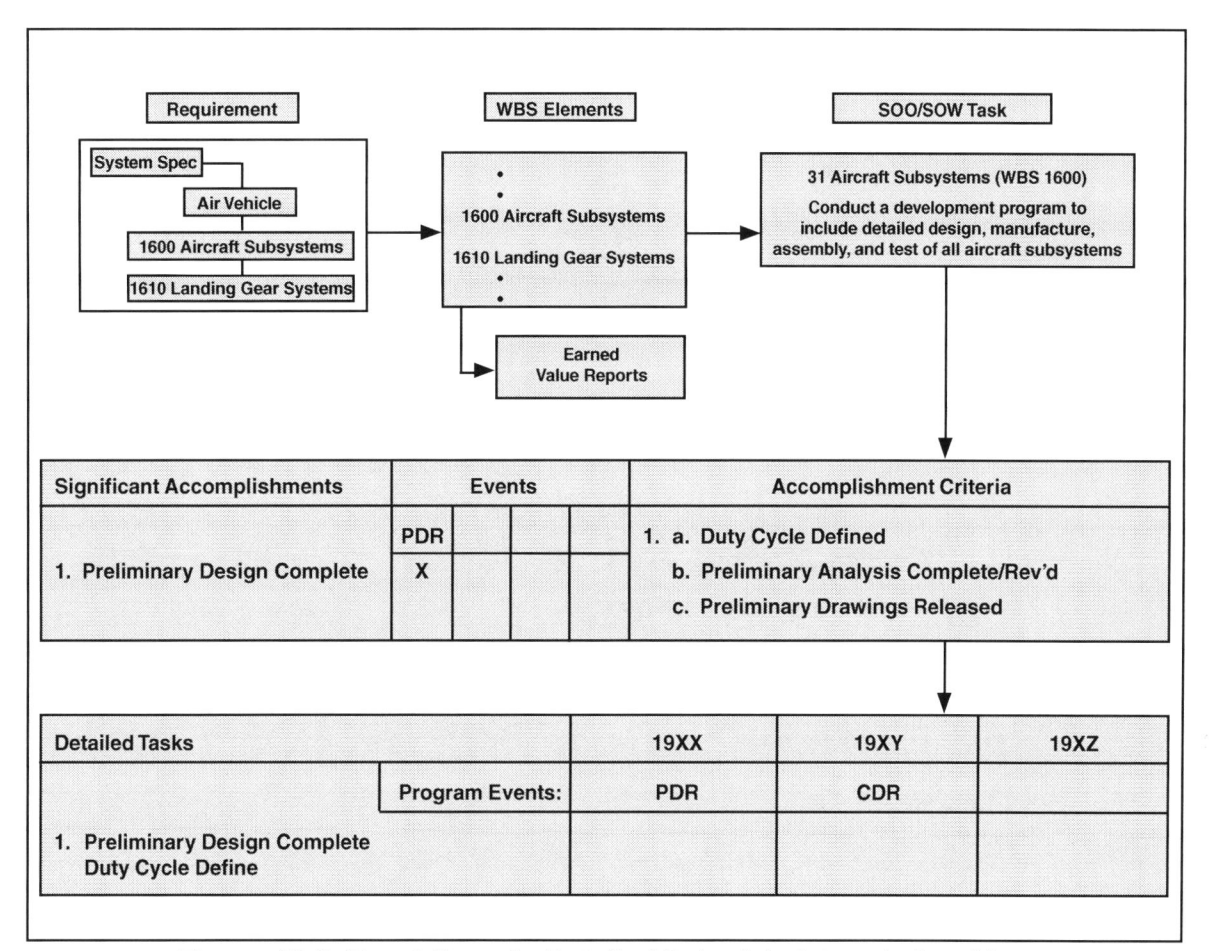

Requirement	WBS Elements	SOO/SOW Task
System Spec → Air Vehicle → 1600 Aircraft Subsystems → 1610 Landing Gear Systems	• 1600 Aircraft Subsystems • 1610 Landing Gear Systems •	31 Aircraft Subsystems (WBS 1600) Conduct a development program to include detailed design, manufacture, assembly, and test of all aircraft subsystems

WBS Elements → Earned Value Reports

Significant Accomplishments	Events			Accomplishment Criteria
	PDR			1. a. Duty Cycle Defined
1. Preliminary Design Complete	X			b. Preliminary Analysis Complete/Rev'd
				c. Preliminary Drawings Released

Detailed Tasks		19XX	19XY	19XZ
	Program Events:	PDR	CDR	
1. Preliminary Design Complete Duty Cycle Define				

Figure 16-3. Event-Based—Detailed Schedule Interrelationships

Schedule Summary

The event-based schedule establishes the key tasks and results expected. The event-based schedule establishes the basis for a valid calendar-based (detail) schedule.

CHAPTER 17

PRODUCT IMPROVEMENT STRATEGIES

17.1 INTRODUCTION

Complex systems do not usually have stagnant configurations. A need for a change during a system's life cycle can come from many sources and effect the configuration in infinite ways. The problem with these changes is that, in most cases it is difficult, if not impossible, to predict the nature and timing of these changes at the beginning of system development. Accordingly, strategies or design approaches have been developed to reduce the risk associated with predicted and unknown changes.

Well thought-out improvement strategies can help control difficult engineering problems related to:

- Requirements that are not completely understood at program start,

- Technology development that will take longer than the majority of the system development,

- Customer needs (such as the need to combat a new military threat) that have increased, been upgraded, are different, or are in flux,

- Requirements change due to modified policy, operational philosophy, logistics support philosophy, or other planning or practices from the eight primary life cycle function groups,

- Technology availability that allows the system to perform better and/or less expensively,

- Potential reliability and maintainability upgrades that make it less expensive to use, maintain, or support, including development of new supply support sources,

- Safety issues requiring replacement of unsafe components, and

- Service life extension programs that refurbish and upgrade systems to increase their service life.

In DoD, the 21st century challenge will be improving existing products and designing new ones that can be easily improved. With the average service life of a weapons system in the area of 40 or more years, it is necessary that systems be developed with an appreciation for future requirements, foreseen and unforeseen. These future requirements will present themselves as needed upgrades to safety, performance, supportability, interface compatibility, or interoperability; changes to reduce cost of ownership; or major rebuild. Providing these needed improvements or corrections form the majority of the systems engineer's post-production activities.

17.2 PRODUCT IMPROVEMENT STRATEGIES

As shown by Figure 17-1, these strategies vary based on where in the life cycle they are applied. The strategies or design approaches that reflect these improvement needs can be categorized as planned improvements, changes in design or production, and deployed system upgrades.

Planned Improvements

Planned improvements strategies include evolutionary acquisition, preplanned product development, and open systems. These strategies are not exclusive and can be combined synergistically in a program development.

Systems Engineering Fundamentals — Chapter 17

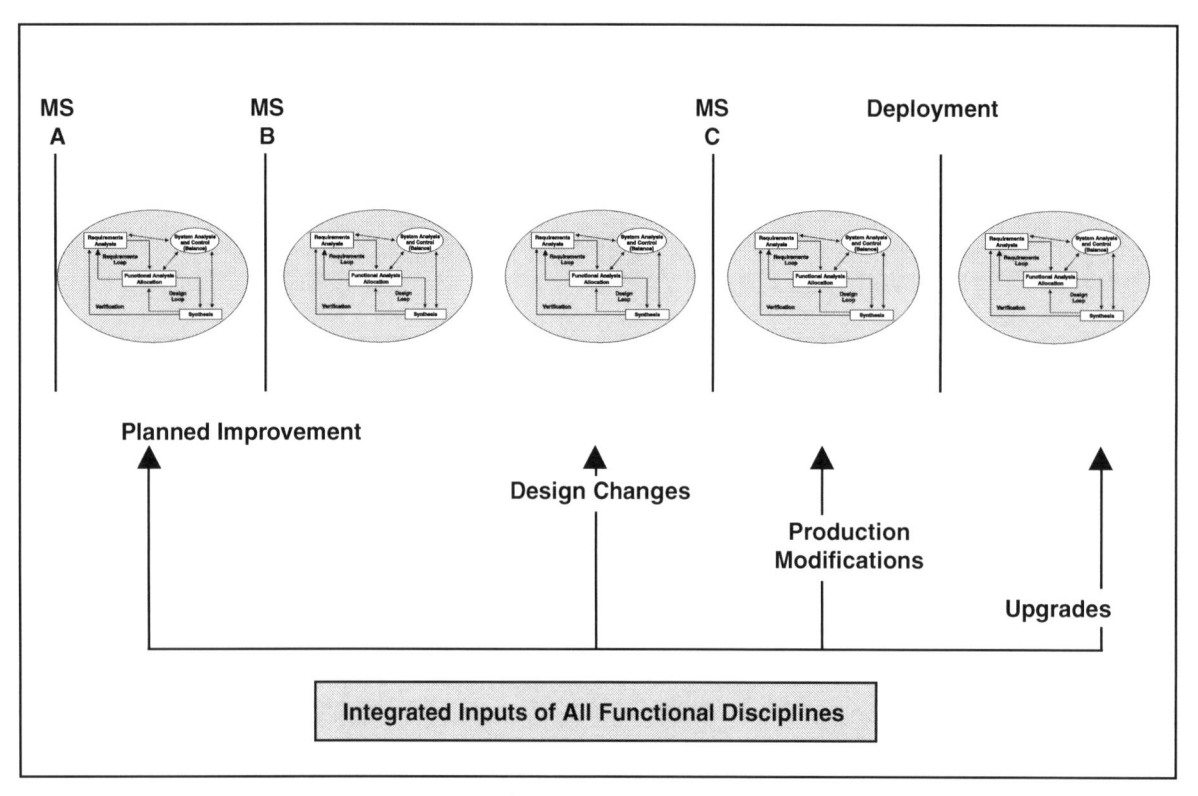

Figure 17-1. Types of Product Improvement Strategies

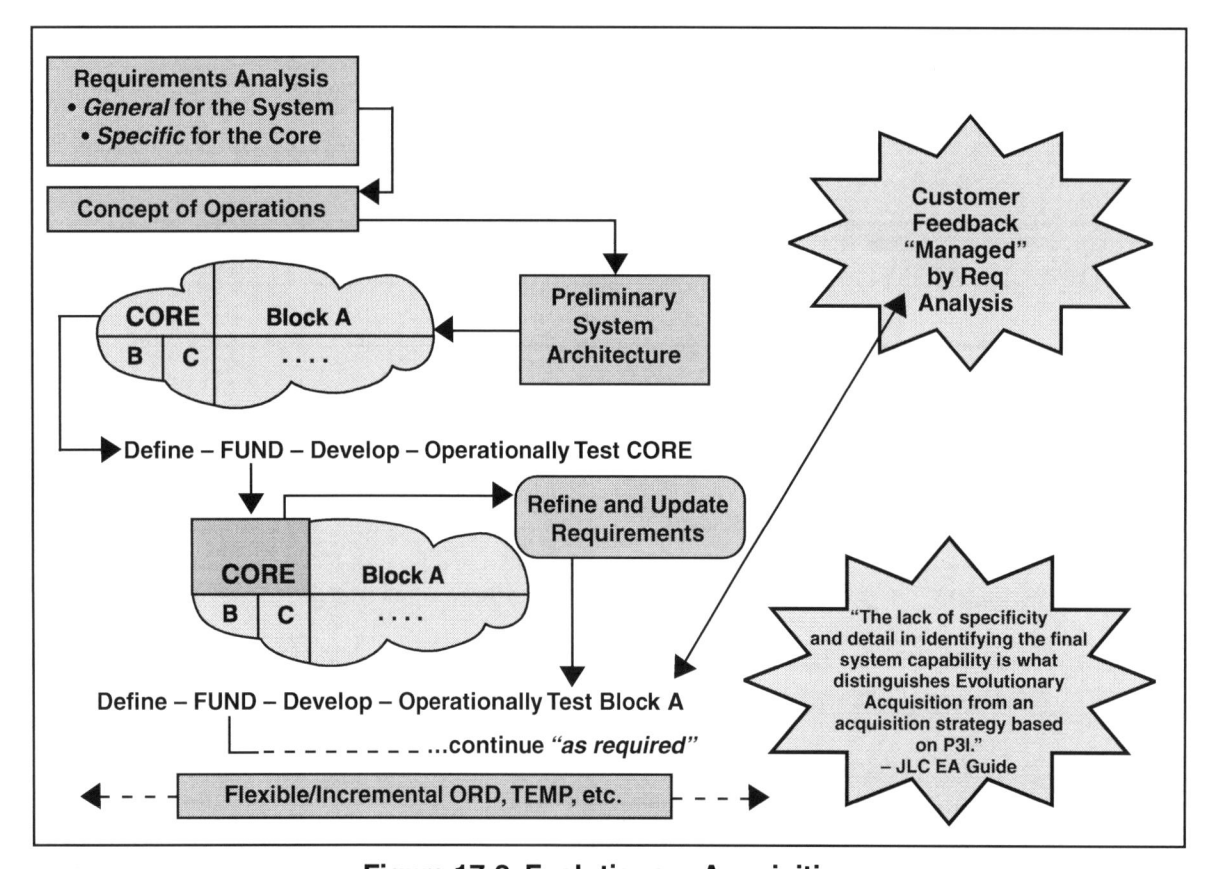

Figure 17-2. Evolutionary Acquisition

Evolutionary Acquisition: Evolutionary acquisition is the preferred approach to systems acquisition in DoD. In an environment where technology is a fast moving target and the key to military superiority is a technically superior force, the requirement is to transition useful capability from development to the user as quickly as possible, while laying the foundation for further changes to occur at later dates. Evolutionary acquisition is an approach that defines requirements for a core capability, with the understanding that the core is to be augmented and built upon (evolved) until the system meets the full spectrum of user requirements. The core capability is defined as a function of user need, technology maturity, threat, and budget. The core is then expanded as need evolves and the other factors mentioned permit.

A key to achieving evolutionary acquisition is the use of time-phased requirements and continuous communication with the eventual user, so that requirements are staged to be satisfied incrementally, rather than in the traditional single grand design approach. Planning for evolutionary acquisition also demands that engineering designs be based on open system, modular design concepts that permit additional increments to be added over time without having to completely re-design and re-develop those portions of the system already fielded. Open designs will facilitate access to recent changes in technologies and will also assist in controlling costs by taking advantage of commercial competition in the marketplace. This concept is not new; it has been employed for years in the C4ISR community, where system are often in evolution over the entire span of their lifecycles.

Preplanned Product Improvement (P3I): Often referred to as P3I, preplanned product improvement is an appropriate strategy when requirements are known and firm, but where constraints (typically either technology or budget) make some portion of the system unachievable within the schedule required. If it is concluded that a militarily

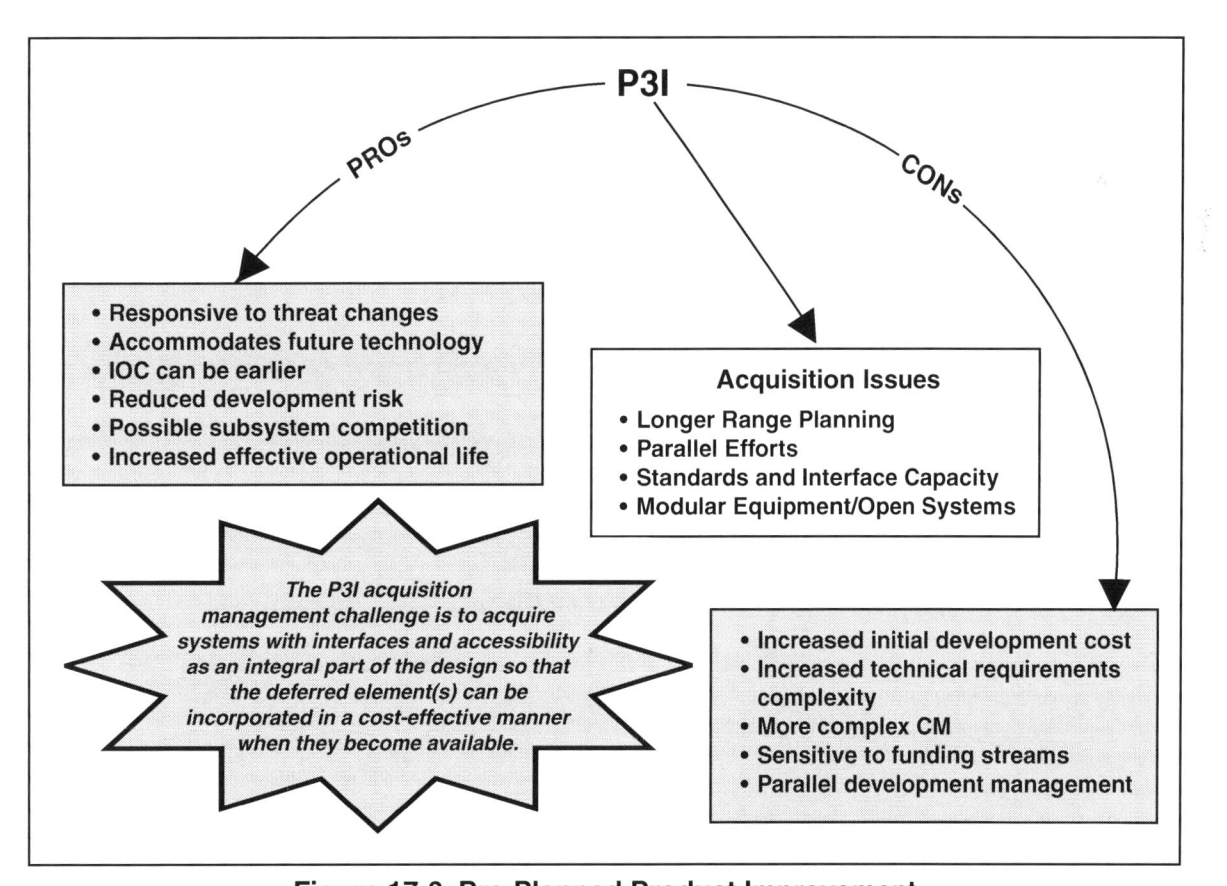

P3I

PROs
- Responsive to threat changes
- Accommodates future technology
- IOC can be earlier
- Reduced development risk
- Possible subsystem competition
- Increased effective operational life

CONs — Acquisition Issues
- Longer Range Planning
- Parallel Efforts
- Standards and Interface Capacity
- Modular Equipment/Open Systems

The P3I acquisition management challenge is to acquire systems with interfaces and accessibility as an integral part of the design so that the deferred element(s) can be incorporated in a cost-effective manner when they become available.

- Increased initial development cost
- Increased technical requirements complexity
- More complex CM
- Sensitive to funding streams
- Parallel development management

Figure 17-3. Pre-Planned Product Improvement

useful capability can be fielded as an interim solution while the portion yet to be proceeds through development, then P3I is appropriate. The approach generally is to handle the improvement as a separate, parallel development; initially test and deliver the system without the improvement; and prove and provide the enhanced capability as it becomes available. The key to a successful P3I is the establishment of well-defined interface requirements for the system and the improvement. Use of a P3I will tend to increase initial cost, configuration management activity, and technical complexity. Figure 17-3 shows some of the considerations in deciding when it is appropriate.

Open Systems Approach: The open system design approach uses interface management to build flexible design interfaces that accommodate use of competitive commercial products and provide enhanced capacity for future change. It can be used to prepare for future needs when technology is yet not available, whether the operational need is known or unknown. The open systems focus is to design the system such that it is easy to modify using standard interfaces, modularity, recognized interface standards, standard components with recognized common interfaces, commercial and nondevelopmental items, and compartmentalized design. Open system approaches to design are further discussed at the end of this chapter.

Changes in Design or Production

Engineering Change Proposals (ECPs): Changes that are to be implemented during the development and production of a given system are typically initiated through the use of ECPs. If the proposed change is approved (usually by a configuration control board) the changes to the documentation that describes the system are handled by formal configuration management, since, by definition, ECPs, when approved, change an approved baseline. ECPs govern the scope and details of these changes. ECPs may address a variety of needs, including correction of deficiencies, cost reduction, and safety. Furthermore, ECPs may been assigned differing levels of priority from routine to emergency. MIL-HDBK-61, Configuration Management Guidance, offers an excellent source of advice on issues related to configuration changes.

Block Change before Deployment: Block changes represent an attempt to improve configuration management by having a number of changes grouped and applied such that they will apply consistently to groups (or blocks) of production items. This improves the management and configuration control of similar items substantially in comparison to change that is implemented item by item and single change order by single change order. When block changes occur, the life cycle impact should be carefully addressed. Significant differences in block configurations can lead to different manuals, supply documentation, training, and restrictions as to locations or activities where the system can be assigned.

Deployed Systems Upgrades

Major Rebuild: A major rebuild results from the need for a system that satisfies requirements significantly different or increased from the existing system, or a need to extend the life of a system that is reaching the end of its usable life. In both cases the system will have upgraded requirements and should be treated as basically a new system development. A new development process should be started to establish and control configuration baselines for the rebuilt system based on the updated requirements.

Major rebuilds include remanufacturing, service-life extension programs, and system developments where significant parts of a previous system will be reused. Though rebuilding existing systems can dramatically reduce the cost of a new system in some cases, the economies of rebuild can be deceiving, and the choice of whether to pursue a rebuild should be done after careful use of trade studies. The key to engineering such systems is to remember that they are new systems and require the full developmental considerations of baselining, the systems engineering process, and life cycle integration.

Post-Production Improvement: In general, product improvements become necessary to improve the system or to maintain the system as its components reach obsolescence. These projects generally result in a capability improvement, but for all practical purposes the system still the serves the same basic need. These improvements are usually characterized by an upgrade to a component or subsystem as opposed to a total system upgrade.

Block Upgrades: Post-production block upgrades are improvements to a specific group of the system population that provides a consistent configuration within that group. Block upgrades in post-production serve the same general purpose of controlling individual system configurations as production block upgrades, and they require the same level of life-cycle integration.

Modifying an Existing System

Upgrading an existing system is a matter of following the system engineering process, with an emphasis on configuration and interface management. The following activities should be included when upgrading a system:

- Benchmark the modified requirements both for the upgrade and the system as a whole,

- Perform functional analysis and allocation on the modified requirements,

- Assess the actual capability of the pre-upgrade system,

- Identify cost and risk factors and monitor them,

- Develop and evaluate modified system alternatives,

- Prototype the chosen improvement alternative, and

- Verify the improvement.

Product improvement requires special attention to configuration and interface management. It is not uncommon that the existing system's configuration will not be consistent with the existing configuration data. Form, fit, and especially function interfaces often represent design constraints that are not always readily apparent at the outset of a system upgrade. Upgrade planning should ensure that the revised components will be compatible at the interfaces. Where interfaces are impacted, broad coordination and agreement is normally required.

Traps in Upgrading Deployed Systems

When upgrading a deployed system pay attention to the following significant traps:

Scheduling to minimize operational impacts: The user's operational commitments will dictate the availability of the system for modification. If the schedule conflicts with an existing or emerging operational need, the system will probably not become available for modification at the time agreed to. Planning and contractual arrangements must be flexible enough to accept unforeseen schedule changes to accommodate user's unanticipated needs.

Configuration and interface management: Configuration management must address three configurations: the actual existing configuration, the modification configuration, and the final system configuration. The key to successful modification is the level of understanding and control associated with the interfaces.

Logistics compatibility problems: Modification will change the configuration, which in most cases will change the supply support and maintenance considerations. Coordination with the logistics community is essential to the long-term operational success of the modification.

Minimal resources available: Modifications tend to be viewed as simple changes. As this chapter has pointed out, they are not; and they should be carefully planned. That planning should include an estimate of needed resources. If the resources are not available, either the project should be abandoned, or a plan formulated to mitigate and control the risk of an initial, minimal budget combined with a plan for obtaining additional resources.

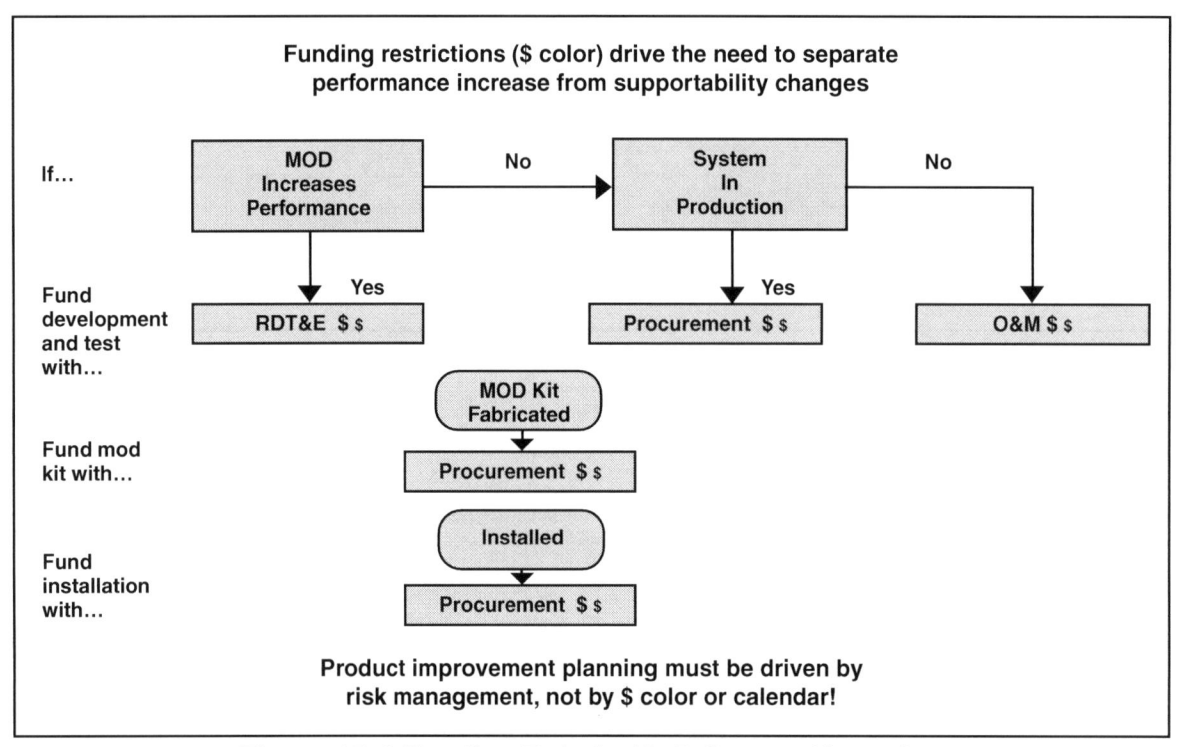

Figure 17-4. Funding Rule for DoD System Upgrades

Funding restrictions ($ color) drive the need to separate performance increase from supportability changes

- If MOD Increases Performance → Yes: Fund development and test with RDT&E $ $
- If No → System In Production → Yes: Procurement $ $
- If No → O&M $ $
- MOD Kit Fabricated → Fund mod kit with Procurement $ $
- Installed → Fund installation with Procurement $ $

Product improvement planning must be driven by risk management, not by $ color or calendar!

Limited competitors: Older systems may have only a few suppliers that have a corporate knowledge of the particular system functions and design. This is especially problematic if the original system components were commercial or NDIs that the designer does not have product baseline data for. In cases such as these, there is a learning process that must take place before the designer or vendor can adequately support the modification effort. Depending on the specific system, this could be a major effort. This issue should be considered very early in the modification process because it has serious cost implications.

Government funding rules: As Figure 17-4 shows the use of government funding to perform system upgrades has restrictions. The purpose of the upgrade must be clear and justified in the planning efforts.

17.3 ROLES AND RESPONSIBILITIES

Modification management is normally a joint government and contractor responsibility. Though any specific system upgrade will have relationships established by the conditions surrounding the particular program, government responsibilities would usually include:

- Providing a clear statement of system requirements,
- Planning related to government functions,
- Managing external interfaces,
- Managing the functional baseline configuration, and
- Verifying that requirements are satisfied.

Contractor responsibilities are established by the contract, but would normally include:

- Technical planning related to execution,
- Defining the new performance envelope,
- Designing and developing modifications, and

- Providing evidence that changes made have modified the system as required.

System Engineering Role

The systems engineering role in product improvement includes:

- Planning for system change,

- Applying the systems engineering process,

- Managing interface changes,

- Identifying and using interface standards which facilitate continuing change,

- Ensuring life cycle management is implemented,

- Monitoring the need for system modifications, and

- Ensuring operations, support activities, and early field results are considered in planning.

17.4 SUMMARY POINTS

- Complex systems do not usually have stagnant configurations.

- Planned improvements strategies include evolutionary acquisition, preplanned product development, and open systems.

- A major rebuild should be treated as a new system development.

- Upgrading an existing system is a matter of following the system engineering process, with an emphasis on configuration and interface management.

- Pay attention to the traps. Upgrade projects have many.

SUPPLEMENT 17-A

OPEN SYSTEM APPROACH

The open system approach is a business and technical approach to system development that results in systems that are easier to change or upgrade by component replacement. It is a system development logic that emphasizes flexible interfaces and maximum interoperability, optimum use of commercial competitive products, and enhanced system capacity for future upgrade. The value of this approach is that open systems have flexibility, and that flexibility translates into benefits that can be recognized from business, management, and technical perspectives.

From a management and business view, the open system approach directs resources to a more intensive design effort with the expectation of a life cycle cost reduction. As a business approach it supports the DoD policy initiatives of CAIV, increased competition, and use of commercial products. It is a technical approach that emphasizes systems engineering, interface control, modular design, and design for upgrade. As a technical approach it supports the engineering goals of design flexibility, risk reduction, configuration control, long-term supportability, and enhanced utility.

Open Systems Initiative

In DoD the open system initiative was begun as a result of dramatic changes in the computer industry that afforded significant advantages to design of C4ISR and IT systems. The standardization achieved by the computer industry allows C4ISR and IT systems to be designed using interface standards to select off-the-shelf components to form the system. This is achieved by using commercially-supported specifications and standards for specifying system interfaces (external and internal, functional and physical), products, practices, and tools. An open system is one

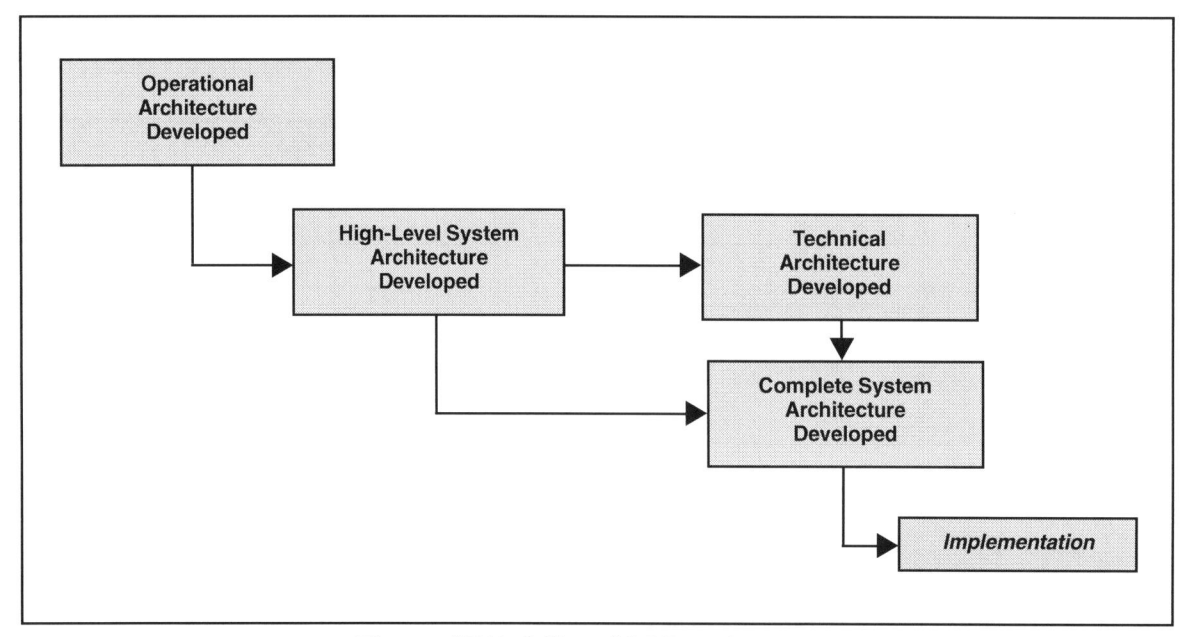

Figure 17-5. C4I and IT Development

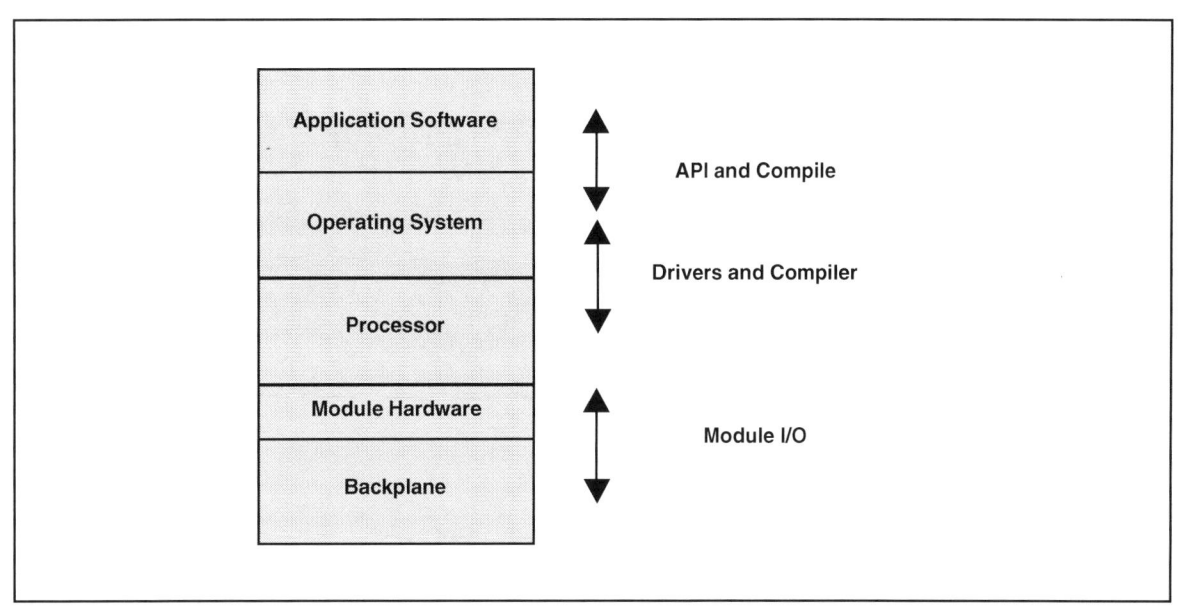

Figure 17-6. Simplified Computer Resource Reference Model

in which interfaces are fully described by open standards.[1] An open system approach extends this concept further by using modular design and interface design to enhance the availability of multiple design solutions, especially those reflecting use of open standards, competitive commercial components, NDIs, and future upgrade capability.

As developed in the C4ISR and IT communities, the open system approach requires the design of three architectures: operational, technical, and system.

As shown in Figure 17-5, the first one prepared is an operational architecture that defines the tasks, operational elements, and information flows required to accomplish or support an operational function. The user community generates the operational concepts that form an operational architecture. The operational architecture is allusive. It is not a specific document required to be developed by the user such as the ORD; but because of their operational nature, the user must provide the components of the operational architecture. It is usually left to the developer to assemble and structure the information as part of the system definition requirements analysis. Once the operational architecture has clearly defined the operational need, development of a system architecture[2] is begun.

The (open) system architecture is a set of descriptions, including graphics, of systems and interconnections supporting the operational functions described in the operational architecture. Early in the (open) system architecture development a technical architecture is prepared to establish a set of rules, derived from open consensus-based industry standards, to govern the arrangement, interaction, and interdependence of the elements of a reference model. Reference models are a common conceptual framework for the type of system being designed. (A simple version for computer resources is shown in Figure 17-6.)

---

[1] Open Standards are non-proprietary, consensus-based standards widely accepted by industry. Examples include SAE, IEEE, and ISO standards.

[2] This system architecture typically describes the end product but not the enabling products. It relies heavily on interface definitions to describe system components.

The technical architecture identifies the services, interfaces, standards, and their relationships; and provides the technical guidelines upon which engineering specifications are based, common building blocks are built, and product lines are developed. In short, the technical architecture becomes a design requirement for developing the system. (The purpose, form, and function of the technical architecture is similar to building codes.)

The system architecture is then further developed to eventually specify component performance and interface requirements. These are then used to select the specific commercial components that form the system under development. This process, called an *implementation*, envisions the production process as consisting primarily of selecting components, conformance (to the interface and performance requirements) management, and assembly, with little or no need for detailed design fabrications.

The process described above has allowed significant achievements in computer-related developments. Other technical fields have also used the open system design approach extensively. (Common examples are the electrical outlets in your home and the tire-to-wheel interface on your car). In most cases the process is not as well defined as it is in the current digital electronics area. A consistent successful use of the open design concept, in and outside the electronics field, requires an understanding of how this process relates to the activities associated with systems engineering management.

Systems Engineering Management

The open system approach impacts all three essential elements of systems engineering management: systems engineering phasing, the systems engineering process, and life cycle considerations. It requires enhanced interface management in the systems engineering process, and requires specific design products be developed prior to engineering-event milestones. The open systems approach is inherently life-cycle friendly. It favorably impacts production and support functions, but it also requires additional effort to assure life-cycle conformance to interface requirements.

Open Systems Products and SE Development Phasing

A system is developed with stepped phases that allow an understanding of the operational need to eventually evolve into a design solution. Though some tailoring of this concept is appropriate, the basic phasing (based on the operational concept preceding the system description, which precedes the preliminary design, which precedes the detailed design) is necessary to coordinate the overall design process and control the requirements flowdown. As shown by Figure 17-7 the open system approach blends well with these development phases.

Concept Studies Phase

The initial detailed operational concept, including operational architectures, should be a user-community output (with some acquisition engineering assistance) produced during the concept exploration phase that emphasizes operational concepts associated with various material solutions. The operational concept is then updated as necessary for each following phase. Analysis of the initial operational concept should be a key element of the operational view output of the system definition phase requirements analysis. An operational architecture developed for supporting the system description should be complete, comprehensive, and clear; and verified to be so at the Alternative Systems Review. If the operational architecture cannot be completed, then a core operational capability must be developed to establish the basis for further development. Where a core capability is used, core requirements should be complete and firm, and the process for adding expanded requirements should be clear and controlled.

System Definition Phase

System interface definitions, such as the technical architecture, and high-level (open) system architecture should be complete in initial form at the

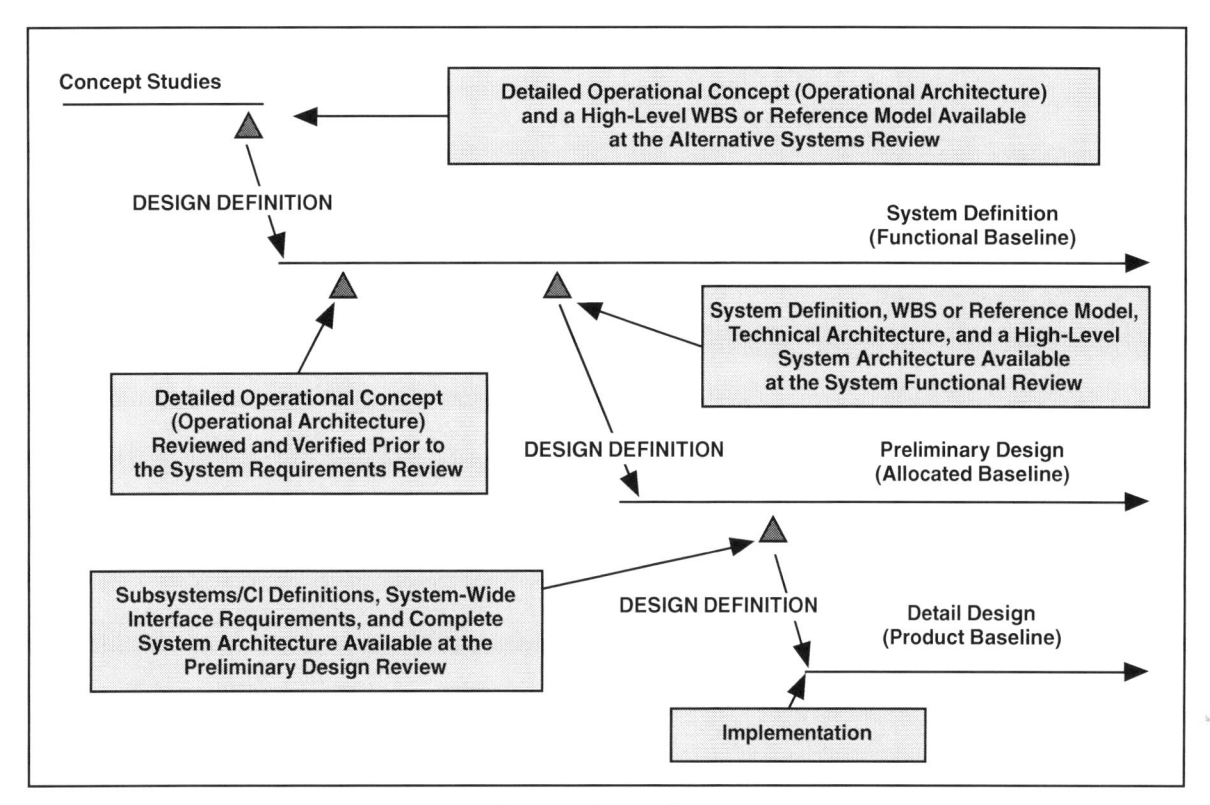

Figure 17-7. Phasing of Open System Development

end of the system definition phase (along with other functional baseline documentation). Successful completion of these items is required to perform the preliminary design, and they should be available for the System Functional Review, also referred to as the System Definition Review or System Design Review. The open system documentation can be separate or incorporated in other functional baseline documentation. The criteria for acceptance should be established in the systems engineering management plan as phase-exit criteria.

Preliminary Design Phase

Along with other allocated baseline documentation, the interface definitions should be updated and the open-system architecture completed by the end of the preliminary design effort. This documentation should also identify the proper level of openness (that is, the level of system decomposition at which the open interfaces are established) to obtain the maximum cost and logistic advantage available from industry practice.

The preliminary design establishes performance-based descriptions of the system components, as well as the interface and structure designs that integrate those components. It is in this phase that the open system approach has the most impact. Interface control should be enhanced and focused on developing modular designs that allow for maximum interchange of competitive commercial products. Review of the technical architecture (or interface definitions) becomes a key element of requirements analysis, open system focused functional partitioning becomes a key element of functional analysis and allocation, iterative analysis of modular designs becomes a key element of design synthesis, and conformance management becomes a key element of verification. Open system related products, such as the technical architecture, interface management documentation, and conformance management documentation, should be key data reviewed at the Preliminary Design Review. Again, the criteria for acceptance should be established in the systems engineering management plan as phase-exit criteria.

Detail Design Phase

The detail design phase becomes the implementation for those parts of the system that have achieved open system status. Conformance management becomes a significant activity as commercial components are chosen to meet performance and interface requirements. Conformance and interface design testing becomes a driving activity during verification to assure an open system or subsystem has been achieved and that components selected meet interface requirements and/or standards.

Systems Engineering Process

The systems engineering problem solving process consists of process steps and loops supported by system analysis and control tools. The focus of the open systems engineering process is compartmentalized design, flexible interfaces, recognized interface standards, standard components with recognized common interfaces, use of commercial and NDIs, and an increased emphasis on interface control. As shown by Figure 17-8, the open-system approach complements the systems engineering process to provide an upgradeable design.

Requirements analysis includes the review and update of interface standards and other interface definitions generated as output from previous systems engineering processes. Functional analysis and allocation focuses on functional partitioning to identify functions that can be performed independent of each other in order to minimize functional interfaces. Design synthesis focuses on modular design with open interfaces, use of open standards compliant commercial products, and the development of performance and interface specifications. The verification processes include conformance testing to validate the interface requirements are appropriate and to verify components chosen to implement the design meet the interface requirements. Engineering open designs, then, does not alter the fundamental practices within systems engineering, but, rather, provides a specific focus to the activities within that process.

System Engineering Control: Interface Management

The key to the open systems engineering process is interface management. Interface management should be done in a more formal and comprehensive manner to rigidly identify all interfaces and

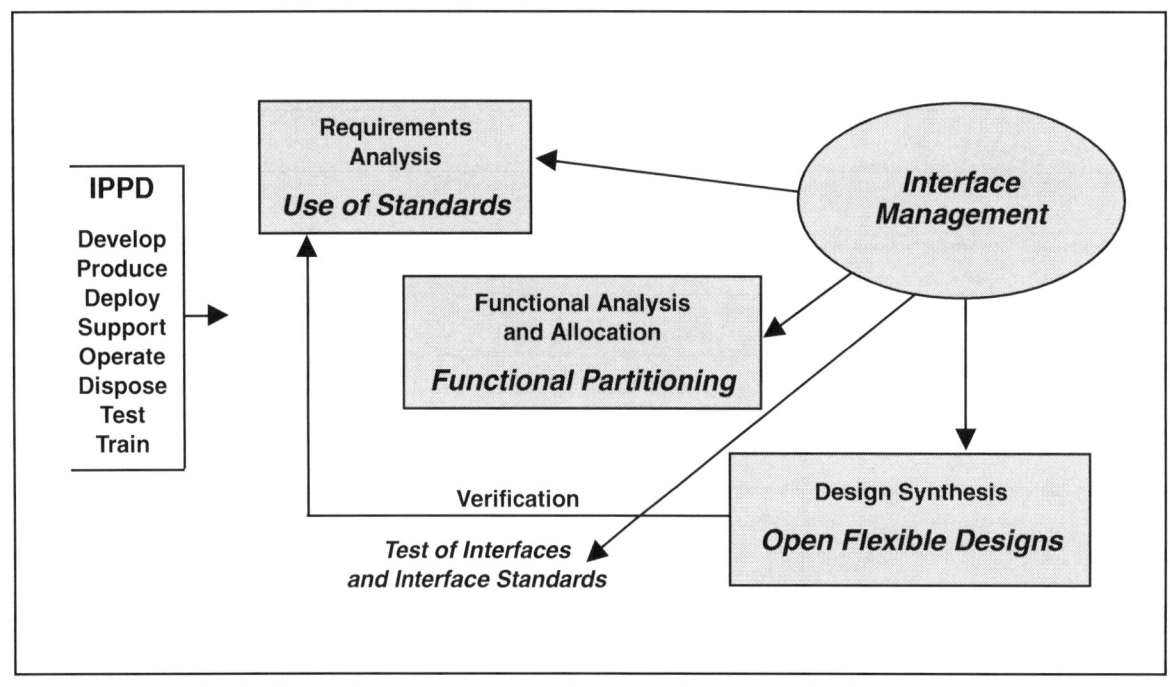

Figure 17-8. Open System Approach to the Systems Engineering Process

control the flowdown and integration of interface requirements. The interfaces become controlled elements of the baseline equal to (or considered part of) the configuration. Open system interface management emphasizes the correlation of interface requirements between interfacing systems. (Do those designing the interfacing systems understand the interface requirements in the same way?) Computer-Aided System Engineering (CASE) generated schematic block diagrams can be used to track interface design activity.

An open system is also characterized by multiple design solutions within the interfaces with emphasis on leveraging best commercial practice. The interface management effort must control interface design such that interfaces specifically chosen for an open system approach are designed based on the following priority:

- Open standards that allow competitive products,

- Open interface design that allows installation of competitive products with minimal change,

- Open interface design that allows minimal change installation of commercial or NDI products currently or planned to be in DoD use, and last,

- Unique design with interfaces designed with upgrade issues considered.

Note that these are clear priorities, not options.

Level of Openness

The level at which the interface design should focus on openness is also a consideration. Each system may have several levels of openness depending on the complexity of the system and the differences in the technology within the system. The level chosen to define the open interfaces should be supported by industry and be consistent with program objectives. For example, for most digital electronics that level is the line-replaceable (LRU) and shop-replaceable (SRU) level. On the other hand the Joint Strike Fighter intends to establish openness at a very high subsystem level to achieve a major program objective, development of different planes using common building blocks (which, in essence, serve as the reference model for the family of aircraft). The open system approach designed segments of a larger system could have additional openness at a lower level. For example, the Advanced Amphibious Assault Vehicle (AAAV) engine compartment is an open approach design allowing for different engine installation and future upgrade capability. On a lower level within the compartment the fuel filters, lines, and connectors are defined by open standard based interfaces. Other systems will define openness at other levels. Program objectives (such as interoperability, upgrade capability, cost-effective support, affordability, and risk reduction) and industry practice (based on market research) drive the choice of the level of openness that will best assure optimum utility and availability of the open system approach.

Life Cycle Considerations

Life cycle integration is established primarily through the use of integrated teaming that combines the design and life cycle planning. The major impacts on life-cycle activity include:

- *Time and cost to upgrade a system is reduced.* It is common in defense systems, which have average life spans in excess of 40 years, that they will require upgrade in their life due to obsolescence of original components, threat increase, and technology push that increases economy or performance. (Most commercial products are designed for a significantly shorter life than military systems, and designs that rely on these commercial products must expect that original commercial components will not necessarily be available throughout the system's life cycle.) By using an open system approach the ability to upgrade a system by changing a single or set of components is greatly enhanced. In addition, the open system approach eases the design problem of replacing the component, thereby reducing the cost and schedule of upgrade, which in turn reduces the operational impact.

- *An open system approach enhances the use of competitive products to support the system.* This flexibility tends to reduce the cost associated with supply support, but more importantly improves component and parts availability.

- *Conformance management becomes a part of the life cycle configuration process.* Replacement of components in an open system must be more controlled because the government has to control the system configuration without controlling the detail component configuration (which will come from multiple sources, all with different detail configurations). The government must expect that commercial suppliers will control the design of their components without regard to the government's systems. The government therefore must use performance- and interface-based specifications to assure the component will provide service equivalent to that approved through the acquisition process. Conformance management is the process that tracks the interface requirements through the life cycle, and assures that the new product meets those requirements.

Summary Comments

Open system design is not only compatible with systems engineering; it represents an approach that enhances the overall systems engineering effort. It controls interfaces comprehensively, provides interface visibility, reduces risk through multiple design solutions, and insists on life cycle interface control. This emphasis on interface identification and control improves systems engineers' capability to integrate the system, probably one of the hardest jobs they have. It also improves the tracking of interface requirements flow down, another key job of the systems engineer. Perhaps most importantly, this rigorous interface management improves systems engineers' ability to correctly determine where commercial items can be properly used.

CHAPTER 18

ORGANIZING AND INTEGRATING SYSTEM DEVELOPMENT

18.1 INTEGRATED DEVELOPMENT

DoD has, for years, required that system designs be integrated to balance the conflicting pressure of competing requirements such as performance, cost, supportability, producibility, and testability. The use of multi-disciplinary teams is the approach that both DoD and industry increasing have taken to achieve integrated designs. Teams have been found to facilitate meeting cost, performance, and other objectives from product concept through disposal.

The use of multi-disciplinary teams in design is known as Integrated Product and Process Development, simultaneous engineering, concurrent engineering, Integrated Product Development, Design-Build, and other proprietary and non-proprietary names expressing the same concept. (The DoD use of the term Integrated Product and Process Development (IPPD) is a wider concept that includes the systems engineering effort as an element. The DoD policy is explained later in this chapter.) Whatever name is used, the fundamental idea involves multi-functional, integrated teams (preferably co-located), that jointly derive requirements and schedules that place equal emphasis on product and process development. The integration requires:

- Inclusion of the eight primary functions in the team(s) involved in the design process,

- Technical process specialties such as quality, risk management, safety, etc., and

- Business processes (usually in an advisory capacity) such as, finance, legal, contracts, and other non-technical support.

Benefits

The expected benefits from team-based integration include:

- Reduced rework in design, manufacturing, planning, tooling, etc.,

- Improved first time quality and reduction of product variability,

- Reduced cost and cycle time,

- Reduced risk,

- Improved operation and support, and

- General improvement in customer satisfaction and product quality throughout its life cycle.

Characteristics

The key attributes that characterize a well integrated effort include:

- Customer focus,

- Concurrent development of products and processes,

- Early and continuous life cycle planning,

- Maximum flexibility for optimization,

- Robust design and improved process capability,

- Event-driven scheduling,

- Multi-disciplinary teamwork,

- Empowerment,
- Seamless management tools, and
- Proactive identification and management of risk.

Organizing for System Development

Most DoD program offices are part of a Program Executive Office (PEO) organization that is usually supported by a functional organization, such as a systems command. Contractors and other government activities provide additional necessary support. Establishing a system development organization requires a network of teams that draw from all these organizations. This network, sometimes referred to as the enterprise, represents the interests of all the stakeholders and provides vertical and horizontal communications.

These integrated teams are structured using the WBS and designed to provide the maximum vertical and horizontal communication during the development process. Figure 18-1 shows how team structuring is usually done. At the system level there is usually a management team and a design team. The management team would normally consist of the government and contractor program managers, the deputy program manager(s), possibly the contractor Chief Executive Officer, the contracting officer, major advisors picked by the program manager, the system design team leader, and other key members of the system design team. The design team usually consists of the first-level subsystem and life-cycle integrated team leaders.

The next level of teams is illustrated on Figure 18-1 as either product or process teams. These teams are responsible for designing system segments (product teams) or designing the supporting or enabling products (process teams). At this level the process teams are coordinating the system level process development. For example, the support team will integrate the supportability analysis from the parts being generated in lower-level design and

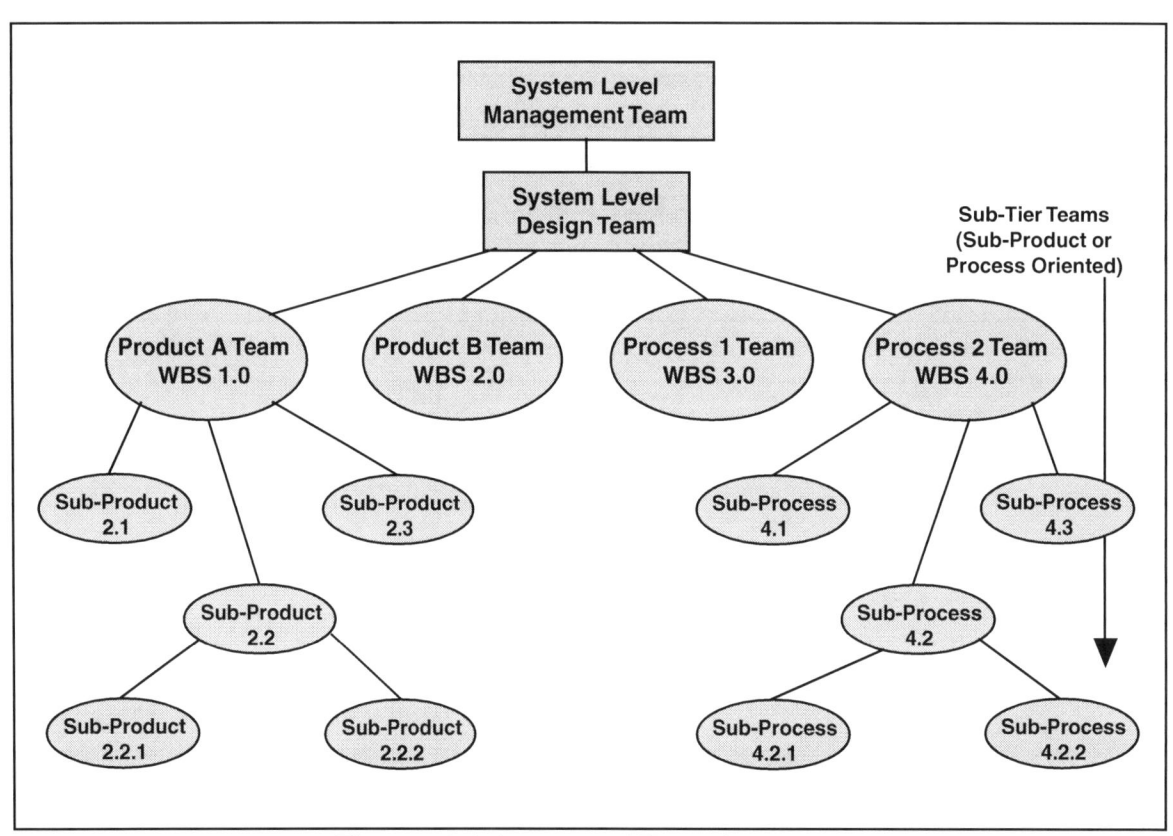

Figure 18-1. Integrated Team Structure

support process teams. Teams below this level continue the process at a lower level of decomposition. Teams are formed only to the lowest level necessary to control the integration. DoD team structures rarely extend lower than levels three or four on the WBS, while contractor teams may extend to lower levels, depending on the complexities of the project and the approach favored by management.

The team structure shown by Figure 18-1 is a hierarchy that allows continuous vertical communication. This is achieved primarily by having the team leaders, and, if appropriate, other key members of a team, be team members of the next highest team. In this manner the decisions of the higher team is immediately distributed and explained to the next team level, and the decisions of the lower teams are presented to the higher team on a regular basis. Through this method decisions of lower-level teams follow the decision making of higher teams, and the higher-level teams' decisions incorporate the concerns of lower-level teams.

The normal method to obtain horizontal communication is shown in Figure 18-2. At least one team member from the Product A Team is also a member of the Integration and Test Team. This member would have a good general knowledge of both testing and Product A. The member's job would be to assist the two teams in designing their end or enabling products, and in making each understand how their decisions would impact the other team. Similarly, the member that sits on both Product A and B teams would have to understand the both technology and the interface issues associated with both items.

The above is an idealized case. Each type of system, each type of contractor organization, and each level of available resources requires a tailoring of this structure. With each phase the focus and the tasks change and so should the structure. As phases

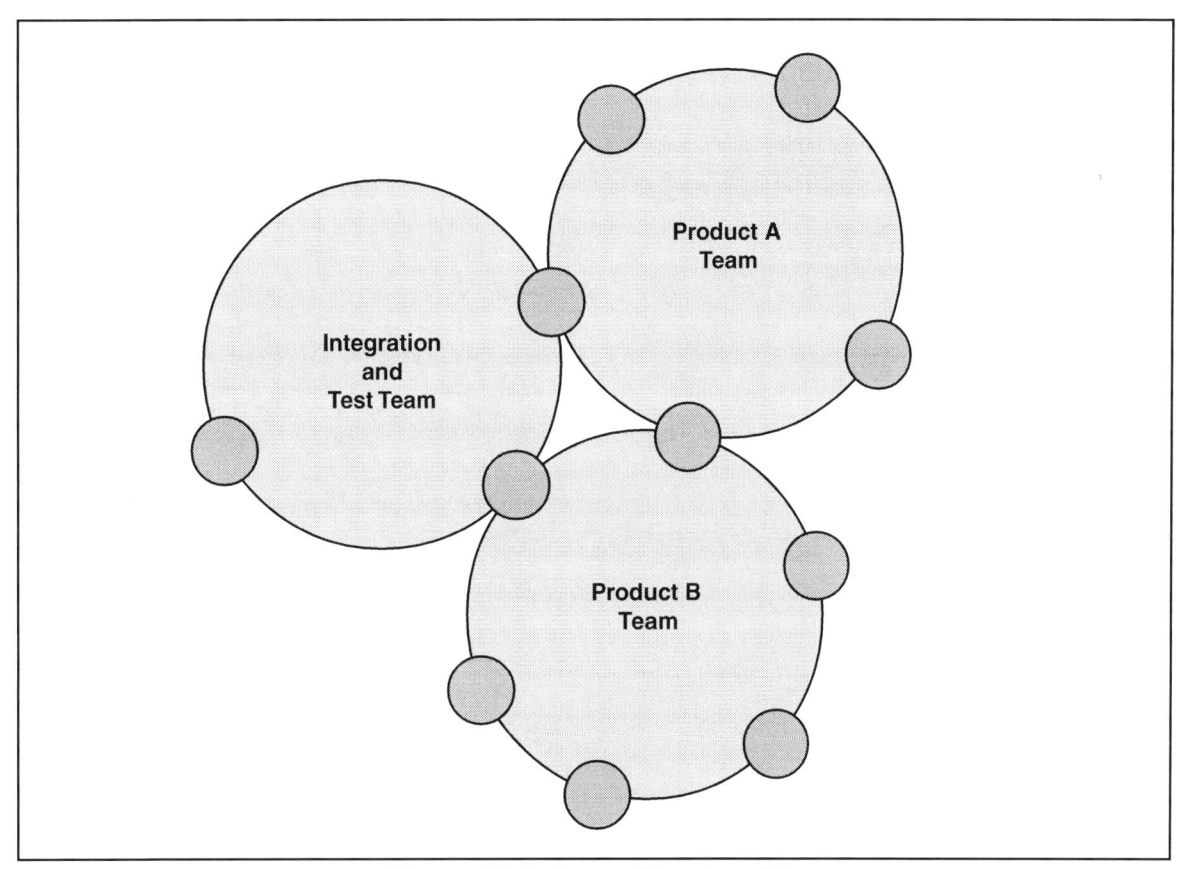

Figure 18-2. Cross Membership

are transited, the enterprise structure and team membership should be re-evaluated and updated.

18.2 INTEGRATED TEAMS

Integrated teams are composed of representatives from all appropriate primary functional disciplines working together with a team leader to:

- Design successful and balanced products,

- Develop the configuration for successful life-cycle control,

- Identify and resolve issues, and

- Make sound and timely decisions.

The teams follow the disciplined approach of the systems engineering process starting with requirements analysis through to the development of configuration baselines as explained earlier in this book. The system-level design team should be responsible for systems engineering management planning and execution. The system-level management team, the highest level program IPT, is responsible for acquisition planning, resource allocation, and management. Lower-level teams are responsible for planning and executing their own processes.

Team Organization

Good teams do not just happen; they are the result of calculated management decisions and actions. Concurrent with development of the enterprise organization discussed above, each team must also be developed. Basically the following are key considerations in planning for a team within an enterprise network:

- The team must have appropriate representation from the primary functions, technical specialties, and business support,

- There must be links to establish vertical and horizontal communication in the enterprise,

- You should limit over-uses of cross membership. Limit membership on three or four teams as a rough rule of thumb for the working level, and

- Ensure appropriate representation of government, contractor, and vendors to assure integration across key organizations.

Team Development

When teams are formed they go through a series of phases before a synergistic self-actuating team is evolved. These phases are commonly referred to as forming, storming, norming and performing. The timing and intensity of each phase will depend on the team size, membership personality, effectiveness of the team building methods employed, and team leadership. The team leaders and an enterprise-level facilitator provide leadership during the team development.

Forming is the phase where the members are introduced to their responsibilities and other members. During this period members will tend to need a structured situation with clarity of purpose and process. If members are directed during this initial phase, their uncertainty and therefore apprehension is reduced. Facilitators controlling the team building should give the members rules and tasks, but gradually reduce the level of direction as the team members begin to relate to each other. As members become more familiar with other members, the rules, and tasks, they become more comfortable in their environment and begin to interact at a higher level.

This starts the storming phase. *Storming* is the conflict brought about by interaction relating to the individuals' manner of dealing with the team tasks and personalities. Its outcome is members who understand the way they have to act with other members to accomplish team objectives. The dynamics of storming can be very complex and intense, making it the critical phase. Some teams will go through it quickly without a visible ripple, others will be loud and hot, and some will never emerge from this phase. The team building facilitators must be alert to dysfunctional activity.

Members may need to be removed or teams reorganized. Facilitators during this period must act as coaches, directing but in a personal collaborative way. They should also be alert for members that are avoiding storming, because the team will not mature if there are members who are not personally committed to participate in it.

Once the team has learned to interact effectively it begins to shape its own processes and become more effective in joint tasks. It is not unusual to see some reoccurrence of storming, but if the storming phase was properly transitioned these incidences should be minor and easily passed. In this phase, *norming*, the team building facilitators become a facilitator to the team—not directing, but asking penetrating questions to focus the members. They also monitor the teams and correct emerging problems.

As the team continues to work together on their focused tasks, their performance improves until they reach a level of self-actuation and quality decision making. This phase, *performing*, can take a while to reach, 18 months to two years for a system-level design team would not be uncommon. During the performing stage, the team building facilitator monitors the teams and corrects emerging problems.

At the start of a project or program effort, team building is commonly done on an enterprise basis with all teams brought together in a team-building exercise. There are two general approaches to the exercise:

- A team-learning process where individuals are given short but focused tasks that emphasize group decision, trust, and the advantages of diversity.

- A group work-related task that is important but achievable, such as a group determination of the enterprise processes, including identifying and removing non-value added traditional processes.

Usually these exercises allow the enterprise to pass through most of the storming phase if done correctly. Three weeks to a month is reasonable for this process, if the members are in the same location. Proximity does matter and the team building and later team performance are typically better if the teams are co-located.

18.3 TEAM MAINTENANCE

Teams can be extremely effective, but they can be fragile. The maintenance of the team structure is related to empowerment, team membership issues, and leadership.

Empowerment

The term empowerment relates to how responsibilities and authority is distributed throughout the enterprise. Maintenance of empowerment is important to promote member ownership of the development process. If members do not have personal ownership of the process, the effectiveness of the team approach is reduced or even neutralized. The quickest way to destroy participant ownership is to direct, or even worse, overturn solutions that are properly the responsibility of the team. The team begins to see that the responsibility for decisions is at a higher level rather than at their level, and their responsibility is to follow orders, not solve problems.

Empowerment requires:

- The flow of authority through the hierarchy of teams, not through personal direction (irrespective of organizational position). Teams should have clear tasking and boundaries established by the higher-level teams.

- Responsibility for decision making to be appropriate for the level of team activity. This requires management and higher-level teams to be specific, clear, complete, and comprehensive in establishing focus and tasking, and in specifying what decisions must be coordinated with higher levels. They should then avoid imposing or overturning decisions more properly in the realm of a lower level.

- Teams at each level be given a clear understanding of their duties and constraints. Within the bounds of those constraints and assigned duties members should have autonomy. Higher-level teams and management either accept their decisions, or renegotiate the understanding of the task.

Membership Issues

Another maintenance item of import is team member turnover. Rotation of members is a fact of life, and a necessary process to avoid teams becoming too closed. However, if the team has too fast a turnover, or new members are not fully assimilated, the team performance level will decline and possibly revert to storming. The induction process should be a team responsibility that includes the immediate use of the new team member in a jointly performed, short term, easily achievable, but important task.

Teams are responsible for their own performance, and therefore should have significant, say over the choice of new members. In addition teams should have the power to remove a member; however, this should be preceded by identification of the problem and active intervention by the facilitator. Removal should be a last resort.

Awards for performance should, where possible, be given to the team rather than individuals (or equally to all individuals on the team). This achieves several things: it establishes a team focus, shows recognition of the team as a cohesive force, recognizes that the quality of individual effort is at least in part due to team influence, reinforces the membership's dedication to team objectives, and avoids team member segregation due to uneven awards. Some variation on this theme is appropriate where different members belong to different organizations, and a common award system does not exist. The system-level management team should address this issue, and where possible assure equitable awards are given team members. A very real constraint on cash awards in DoD rises in the case of teams that include both civilian and military members. Military members cannot be given cash awards, while civilians can. Con-sequently, managers must actively seek ways to reward all team members appropriately, leaving no group out at the expense of others.

Leadership

Leadership is provided primarily by the organizational authority responsible for the program, the enterprise facilitator, and the team leaders. In a DoD program, the organizational leaders are usually the program manager and contractor senior manager. These leaders set the tone of the enterprise adherence to empowerment, the focus of the technical effort, and the team leadership of the system management team. These leaders are responsible to see that the team environment is maintained. They should coordinate their action closely with the facilitator.

Facilitators

Enterprises that have at least one facilitator find that team and enterprise performance is easier to maintain. The facilitator guides the enterprise through the team building process, monitors the team network through metrics and other feedback, and makes necessary corrections through facilitation. The facilitator position can be:

- A separate position in the contractor organization,

- Part of the responsibilities of the government systems engineer or contractor project manager, or

- Any responsible position in the first level below the above that is related to risk management.

Obviously the most effective position would be one that allows the facilitator to concentrate on the teams' performance. Enterprise level facilitators should have advanced facilitator training and (recommended) at least a year of mentored experience. Facilitators should also have significant broad experience in the technical area related to the development.

Team Leaders

The team leaders are essential for providing and guiding the team focus, providing vertical communication to the next level, and monitoring the team's performance. Team leaders must have a clear picture of what constitutes good performance for their team. They are not supervisors, though in some organizations they may have supervisory administrative duties. The leader's primary purpose is to assure that the environment is present that allows the team to perform at its optimum level—not to direct or supervise.

The team leader's role includes several difficult responsibilities:

- Taking on the role of coach as the team forms,

- Facilitating as the team becomes self-sustaining,

- Sometimes serving as director (only when a team has failed, needs refocus or correction, and is done with the facilitator),

- Providing education and training for members,

- Facilitating team learning,

- Representing the team to upper management and the next higher-level team, and

- Facilitating team disputes.

Team leaders should be trained in basic facilitator principles. This training can be done in about a week, and there are numerous training facilities or companies that can offer it.

18.4 TEAM PROCESSES

Teams develop their processes from the principles of system engineering management as presented earlier in the book. The output of the teams is the design documentation associated with products identified on the system architecture, including both end product components and enabling products.

Teams use several tools to enhance their productivity and improve communication among enterprise members. Some examples are:

- Constructive modeling (CAD/CAE/CAM/CASE) to enhance design understanding and control,

- Trade-off studies and prioritization,

- Event-driven schedules,

- Prototyping,

- Metrics, and most of all

- Integrated membership that represents the life cycle stakeholders.

Integrated Team Rules

The following is a set of general rules that should guide the activities and priorities of teams in a system design environment:

- Design results must be communicated clearly, effectively, and timely.

- Design results must be compatible with initially defined requirements.

- Continuous "up-the-line" communication must be institutionalized.

- Each member needs to be familiar with all system requirements.

- Everyone involved in the team must work from the same database.

- Only one member of the team has the authority to make changes to one set of master documentation.

- All members have the same level of authority (one person, one vote).

- Team participation is consistent, success-oriented, and proactive.

- Team discussions are open with no secrets.

- Team member disagreements must be reasoned disagreement (alternative plan of action versus unyielding opposition).

- Trade studies and other analysis techniques are used to resolve issues.

- Issues are raised and resolved early.

- Complaints about the team are not voiced outside the team. Conflicts must be resolved internally.

Guidelines for Meeting Management

Even if a team is co-located as a work unit, regular meetings will be necessary. These meetings and their proper running become even more important if the team is not co-located and the meeting is the primary means of one-on-one contact. A well-run technical meeting should incorporate the following considerations:

- Meetings should be held only for a specific purpose and a projected duration should be targeted.

- Advance notice of meetings should normally be at least two weeks to allow preparation and communication between members.

- Agendas, including time allocations for topics and supportive material should be distributed no less than three business days before the team meeting. The objective of the meeting should be clearly defined.

- Stick to the agenda during the meeting. Then cover new business. Then review action items.

- Meeting summaries should record attendance, document any decision or agreements reached, document action items and associated due-dates, provide a draft agenda for the next meeting, and frame issues for higher-level resolution.

- Draft meeting summaries should be provided to members within one working day of the meeting. A final summary should be issued within two working days after the draft comments deadline.

18.5 BARRIERS TO INTEGRATION

There are numerous barriers to building and maintaining a well functioning team organization, and they are difficult to overcome. Any one of these barriers can negate the effectiveness of an integrated development approach. Common barriers include:

- Lack of top management support,

- Team members not empowered,

- Lack of access to a common database,

- Lack of commitment to a cultural change,

- Functional organization not fully integrated into a team process,

- Lack of planning for team effort,

- Staffing requirements conflict with teams,

- Team members not collocated,

- Insufficient team education and training,

- Lessons learned and successful practices not shared across teams,

- Inequality of team members,

- Lack of commitment based on perceived uncertainty,

- Inadequate resources, and

- Lack of required expertise on either the part of the contractor or government.

Breaking Barriers

Common methods to combat barriers include:

- Education and training, and then more education and training: it breaks down the uncertainty of change, and provides a vision and method for success.

- Use a facilitator not only to build and maintain teams, but also to observe and advise management.

- Obtain management support up front. Management must show leadership by managing the teams' environment rather than trying to manage people.

- Use a common database open to all enterprise members.

- Establish a network of teams that integrates the design and provides horizontal and vertical communication.

- Establish a network that does not over-tax available resources. Where a competence is not available in the associated organizations, hire it through a support contractor.

- Where co-location is not possible have regular working sessions of several days duration. Telecommunications, video conferencing, and other technology based techniques can also go far to alleviate the problems of non-collocation.

Summary Comments

- Integrating system development is a systems engineering approach that integrates all essential primary function activities through the use of multi-disciplinary teams, to optimize the design, manufacturing and supportability processes.

- Team building goes through four phases: forming, storming, norming, and performing.

- Key leadership positions in a program network of teams are the program manager, facilitator, and team leaders.

- A team organization is difficult to build and maintain. It requires management attention and commitment over the duration of the teams involved.

SUPPLEMENT 18-A

IPPD – A DOD MANAGEMENT PROCESS

The DoD policy of Integrated Product and Process Development (IPPD) is a broad view of integrated system development which includes not only systems engineering, but other areas involved in formal decision making related to system development. DoD policy emphasizes integrated management at and above the Program Manager (PM) level. It requires IPPD at the systems engineering level, but does not direct specific organizational structures or procedures in recognition of the need to design a tailored IPPD process to every individual situation.

Integrated Product Teams

One of the key IPPD tenets is multi-disciplinary integration and teamwork achieved through the use of Integrated Product Teams (IPTs). While IPTs may not be the best solution for every management situation, the requirement to produce integrated designs that give consideration to a wide array of technical and business concerns leads most organizations to conclude that IPTs are the best organizational approach to systems management. PMs should remember that the participation of a contractor or a prospective contractor on a IPT should be in accordance with statutory requirements, such as procurement integrity rules. The service component's legal advisor must review prospective contractor involvement on IPTs. To illustrate issues the government-contractor team arrangement raises, the text box at the end of this section lists nine rules developed for government members of the Advanced Amphibious Assault Vehicle (AAAV) design IPTs.

The Secretary of Defense has directed that DoD perform oversight and review by using IPTs. These IPTs function in a spirit of teamwork with participants empowered and authorized, to the maximum extent possible, to make commitments for the organization or the functional area they represent. IPTs are composed of representatives from all appropriate functional disciplines working together to build successful programs and enabling decision makers to make the right decisions at the right time.

DoD IPT Structure

The DoD oversight function is accomplished through a hierarchy of teams that include levels of management from DoD to the program level. There are three basic levels of IPTs: the Overaching IPT (OIPT), the Working IPTs (WIPT), and Program IPTs with the focus and responsibilities as shown by Figure 18-3. For each ACAT I program, there will be an OIPT and at least one WIPT. WIPTs will be developed for particular functional topics, e.g., test, cost/performance, contracting, etc. An Integrating IPT (IIPT) will coordinate WIPT efforts and cover all topics not otherwise assigned to another IPT. These teams are structurally organized as shown on Figure 18-4.

Overarching IPT (OIPT)

The OIPT is a DoD level team whose primary responsibility is to advise the Defense Acquisition Executive on issues related to programs managed at that level. The OIPT membership is made up of the principals that are charged with responsibility for the many functional offices at the Office of the Secretary of Defense (OSD).

The OIPT provides:

- Top-level strategic guidance,

Organization	Teams	Focus	Participant Responsibilities
OSD and Components	OIPT*	• Strategic Guidance • Tailoring • Program Assessment • Resolve Issues Elevated by WIPTs	• Program Success • Functional Area Leadership • Independent Assessment • Issue Resolution
	WIPTs*	• Planning for Program Success • Opportunities for Acquisition Reform (e.g. innovation, streamlining) • Identify/Resolve Program Issues • Program Status	• Functional Knowledge and Experience • Empowered Contribution • Recom.'s for Program Success • Communicate Status and Unresolved Issues
Program Teams and System Contractors	Program IPTs**	• Program Execution • Identify and Implement Acquisition Reform	• Manage Complete Scope of Program Resources, and Risk • Integrate Government and Contractor Efforts for Report Program Status and Issues

* Covered in "Rules of the Road"
** Covered in "Guide to Implementation and Management of IPPD in DoD Acquisition"

Figure 18-3. Focus and Responsibilities of IPTs

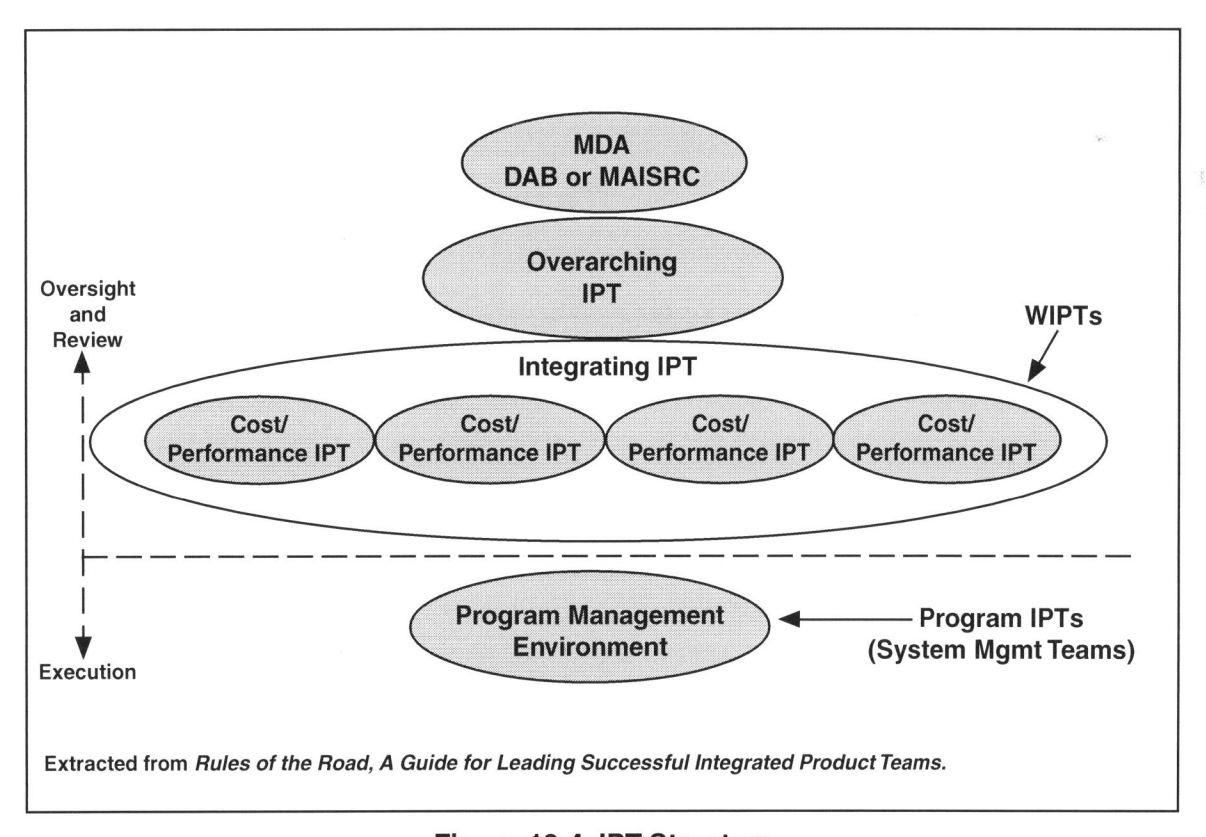

Figure 18-4. IPT Structure

Extracted from *Rules of the Road, A Guide for Leading Successful Integrated Product Teams.*

- Functional area leadership,

- Forum for issue resolution,

- Independent assessment to the MDA,

- Determine decision information for next milestone review, and

- Provide approval of the WIPT structures and resources.

Working-Level IPT (WIPT)

The WIPTs may be thought of as teams that link the PM to the OIPT. WIPTs are typically functionally specialized teams (test, cost-performance, etc.). The PM is the designated head of the WIPT, and membership typically includes representation from various levels from the program to OSD staff. The principal functions of the WIPT are to advise the PM is the area of specialization and to advise the OIPT of program status.

The duties of the WIPT include:

- Assisting the PM in developing strategies and in program planning, as requested by the PM,

- Establishing IPT plan of action and milestones,

- Proposing tailored document and milestone requirements,

- Reviewing and providing early input to documents,

- Coordinating WIPT activities with the OIPT members,

- Resolving or evaluating issues in a timely manner, and

- Obtaining principals' concurrence with applicable documents or portions of documents.

Program IPTs

Program IPTs are teams that perform the program tasks. The integration of contractors with the government on issues relative to a given program truly occurs at the program IPT level. The development teams (product and process teams) described earlier in this chapter would be considered program IPTs. Program IPTs would also include teams formed for business reasons, for example teams established to prepare Planning, Programming, and Budgeting System (PPBS) documentation, to prepare for Milestone Approval, to develop the RFP, or the like.

SUPPLEMENT 18-B

GOVERNMENT ROLE ON IPTs

The following list was developed by the Advanced Amphibious Assault Vehicle (AAAV) program to inform its government personnel of their role on contractor/government integrated teams. It addresses government responsibilities and the realities imposed by contractual and legal constraints. Though it is specific to the AAAV case, it can be used as guidance in the development of team planning for other programs.

1. The IPTs are contractor-run entities. We do not lead or manage the IPTs.

2. We serve as "customer" representatives on the IPTs. We are there to REDUCE THE CYCLE TIME of contractor-Government (customer) communication. In other words, we facilitate contractor personnel getting Government input faster. Government IPT members also enable us to provide the contractor IPT Status and issue information up the Government chain on a daily basis (instead of monthly or quarterly).

3. WE DO NOT DO the contractor's IPT WORK, or any portion of their work or tasks. The contractor has been contracted to perform the tasks outlined in the contract SOW; their personnel and their subcontractors' personnel will perform those tasks, not us. But Government IPT members will be an active part of the deliberations during the development of, and participate in "on-the-fly" reviews of deliverables called out in CDRLs.

4. When asked by contractor personnel for the Government's position or interpretation, Government IPT members can offer their personal opinion, as an IPT member, or offer expert opinion; you can provide guidance as to our "customer" opinion and what might be acceptable to the Government but you can only offer the "Government" position for items that have been agreed to by you and your Supervisor. IT IS UP TO YOUR SUPERVISORS TO EMPOWER EACH OF YOU TO AN APPROPRIATE LEVEL OF AUTHORITY. It is expected that this will start at a minimal level of authority and be expanded as each individual's IPT experience and program knowledge grows. However... (see items 5 and 6).

5. Government IPT members CAN NOT authorize any changes or deviations to/from the contract SOW or Specifications. Government IPT members can participate in the deliberations and discussions that would result in the suggestion of such changes. If/When an IPT concludes that the best course of action is not in accordance with the contract, and a contract change is in order, then the contractor must submit a Contract Change Request (CCR) through normal channels.

6. Government IPT members CAN NOT authorize the contractor to perform work that is in addition to the SOW/contract requirements. The contractor IPTs can perform work that is not specifically required by the contract, at their discretion (provided they stay within the resources as identified in the Team Operating Contract (TOC).

7. Government IPT member participation in contractor IPT activities IS NOT Government consent that the work is approved by the Government or is chargeable to the contract. If an IPT is doing something questionable, identify it to your supervisor or Program Management Team (PMT) member.

8. Government members of IPTs do not approve or disapprove of IPT decisions, plans, or reports. You offer your opinion in their development, you vote as a member, and you coordinate issues with your Supervisor and bring the "Government" opinion (in the form of your opinion) back to the IPT, with the goal of improving the quality of the products; you don't have veto power.

9. Government IPT members are still subject to all the Government laws and regulations regarding "directed changes," ethics, and conduct. Your primary function is to perform those functions that are best done by Government employees, such as:

- Conveying to contractor personnel your knowledge/expertise on Marine Corps operations and maintenance techniques;

- Interfacing with all other Government organizations (e.g., T&E);

- Control/facilitization of government furnished equipment and materials (GFE and GFM);

- Ensuring timely payment of submitted vouchers; and

- Full participation in Risk Management.

CHAPTER 19

CONTRACTUAL CONSIDERATIONS

19.1 INTRODUCTION

This chapter describes how the systems engineer supports the development and maintenance of the agreement between the project office and the contractor that will perform or manage the detail work to achieve the program objectives. This agreement has to satisfy several stakeholders and requires coordination between responsible technical, managerial, financial, contractual, and legal personnel. It requires a document that conforms to the Federal Acquisition Regulations (and supplements), program PPBS documentation, and the System Architecture. As shown by Figure 19-1, it also has to result in a viable cooperative environment that allows necessary integrated teaming to take place.

The role of technical managers or systems engineers is crucial to satisfying these diverse concerns. Their primary responsibilities include:

- Supporting or initiating the planning effort. The technical risk drives the schedule and cost risks which in turn should drive the type of contractual approach chosen,

- Prepares or supports the preparation of the source selection plan and solicitation clauses concerning proposal requirements and selection criteria,

- Prepares task statements,

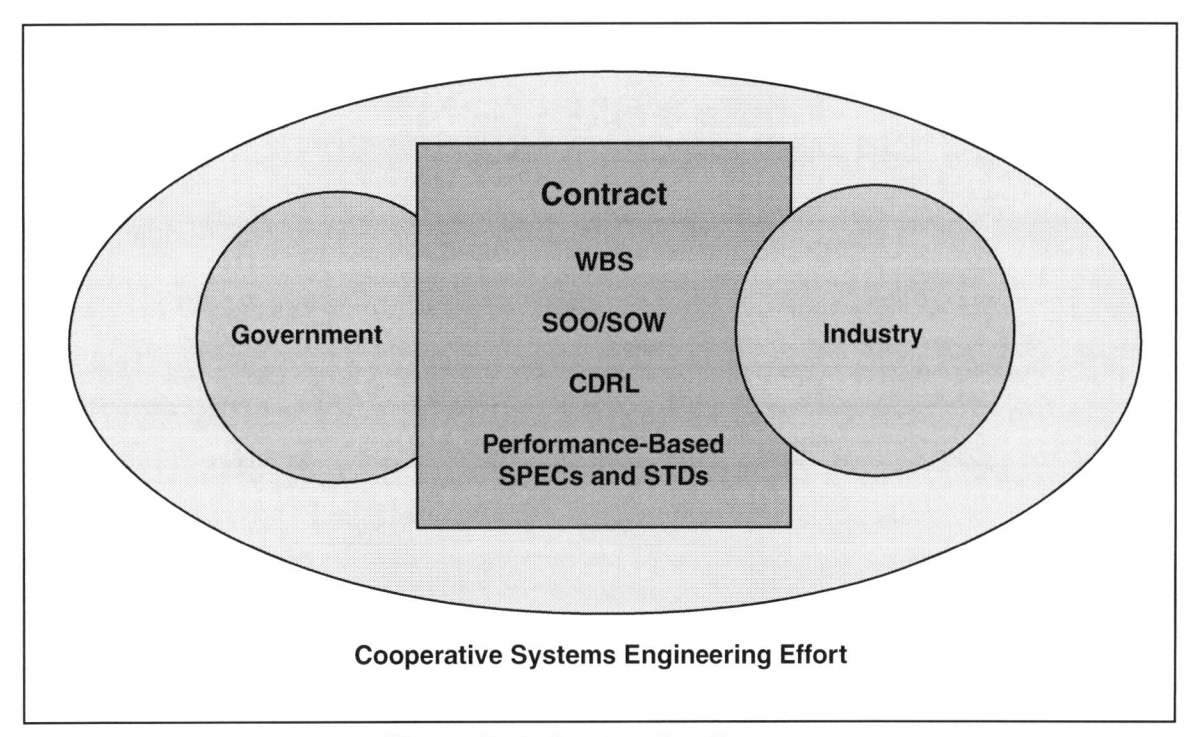

Figure 19-1. Contracting Process

- Prepares the Contract Data Requirements List (CDRL),

- Supports negotiation and participates in source selection evaluations,

- Forms Integrated Teams and coordinates the government side of combined government and industry integrated teams,

- Monitors the contractor's progress, and

- Coordinates government action in support of the contracting officer.

This chapter reflects the DoD approach to contracting for system development. It assumes that there is a government program or project office that is tasking a prime contractor in a competitive environment. However, in DoD there is variation to this theme. Some project activities are tasked directly to a government agency or facility, or are contracted sole source. The processes described in this chapter should be tailored as appropriate for these situations.

19.2 SOLICITATION DEVELOPMENT

As shown by Figure 19-2, the DoD contracting process begins with planning efforts. Planning includes development of a Request for Proposal (RFP), specifications, a Statement of Objective (SOO) or Statement of Work (SOW), a source selection plan, and the Contract Data Requirements List (CDRL).

Request for Proposal (RFP)

The RFP is the solicitation for proposals. The government distributes it to potential contractors. It describes the government's need and what the offeror must do to be considered for the contract. It establishes the basis for the contract to follow.

The key systems engineering documents included in a solicitation are:

- A statement of the work to be performed. In DoD this is a SOW. A SOO can be used to obtain a SOW or equivalent during the selection process.

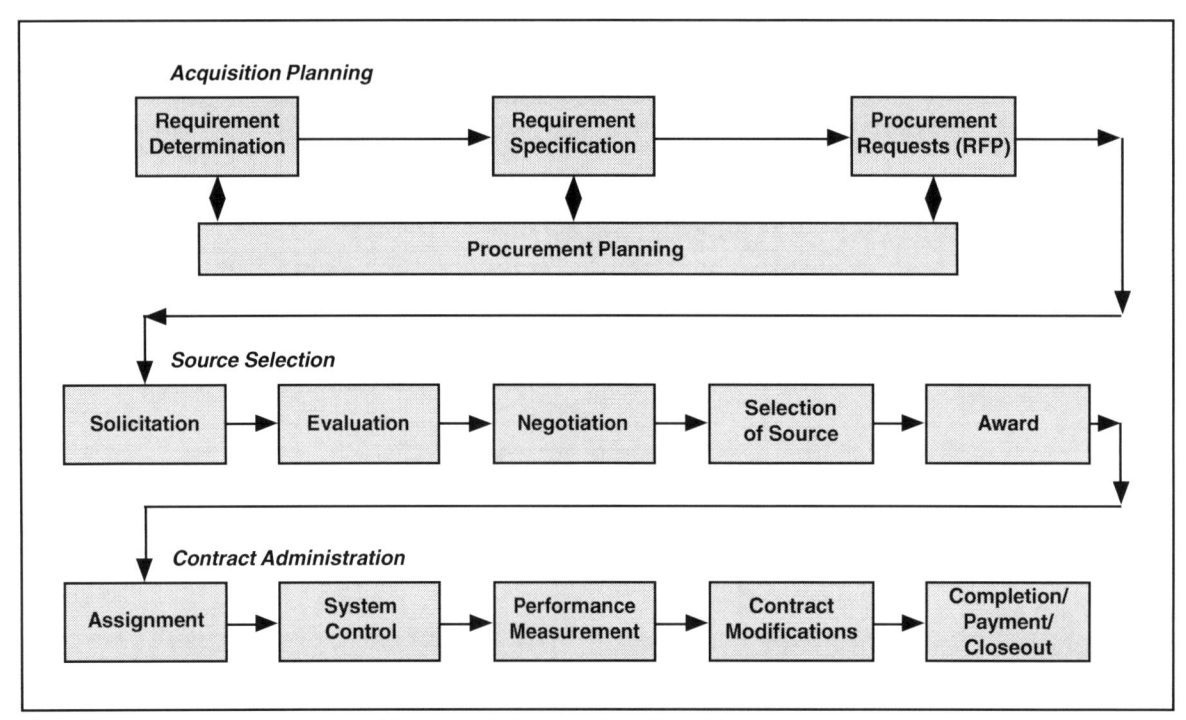

Figure 19-2. Contracting Process

- A definition of the system. Appropriate specifications and any additional baseline information necessary for clarification form this documentation. This is generated by the systems engineering process as explained earlier in this book.

- A definition of all data required by the customer. In DoD this accomplished through use of the Contract Data Requirements List (CDRL).

The information required to be in the proposals responding to the solicitation is also key for the systems engineer. An engineering team will decide the technical and technical management merits of the proposals. If the directions to the offerors are not clearly and correctly stated, the proposal will not contain the information needed to evaluate the offerors. In DoD Sections L and M of the RFP are those pivotal documents.

Task Statement

The task statement prepared for the solicitation will govern what is actually received by the government, and establish criteria for judging contractor performance. Task requirements are expressed in the SOW. During the solicitation phase the tasks can be defined in very general way by a SOO. Specific details concerning SOOs and SOWs are attached at the end of this chapter.

As shown by Figure 19-3, solicitation tasking approaches can be categorized into four basic options: use of a basic operational need, a SOO, a SOW, or a detail specification.

Option 1 maximizes contractor flexibility by submitting the Operational Requirements Document (ORD) to offerors as a requirements document (e.g. in place of SOO/SOW), and the offerors are requested to propose a method of developing a solution to the ORD. The government identifies its areas of concern in Section M (evaluation factors) of the RFP to provide guidance. Section L (instructions to the offerors) should require the bidders write a SOW based on the ORD as *part* of their proposal. The offeror proposes the type of system. The contractor develops the system specification and the Work Breakdown Structure (WBS). In general this option is appropriate for early efforts where contractor input is necessary to expand the understanding of physical solutions and alternative system approaches.

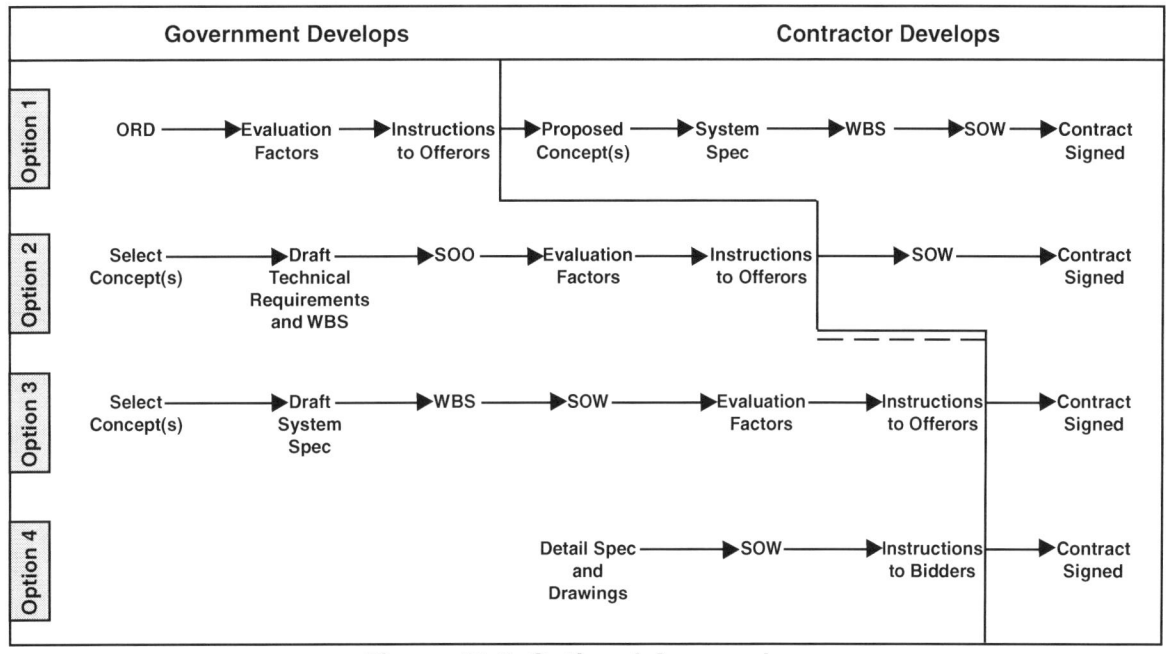

Figure 19-3. Optional Approaches

Option 2 provides moderate contractor flexibility by submitting a SOO to the offerors as the Section C task document (e.g., in place of SOW.) The government identifies its areas of concern in Section M (evaluation factors) to provide guidance. Section L (instructions to the offerors) should require as part of the proposal that offerors write a SOW based on the SOO. In this case the government usually selects the type of system, writes a draft technical-requirements document or system specification, and writes a draft WBS. This option is most appropriate when previous efforts have not defined the system tightly. The effort should not have any significant design input from the previous phase. This method allows for innovative thinking by the bidders in the proposal stage. It is a preferred method for design contracts.

Option 3 lowers contractor flexibility, and increases clarity of contract requirements. In this option the SOW is provided to the Contractor as the contractual task requirements document. The government provides instructions in Section L to the offerors to describe the information needed by the government to evaluate the contractor's ability to accomplish the SOW tasks. The government identifies evaluation factors in Section M to provide guidance for priority of the solicitation requirements. In most cases, the government selects the type of system, and provides the draft system spec, as well as the draft WBS. This option is most appropriate when previous efforts have defined the system to the lower WBS levels or where the product baseline defines the system. Specifically when there is substantial input from the previous design phase and there is a potential for a different contractor on the new task, the SOW method is appropriate.

Option 4 minimizes contractor flexibility, and requires maximum clarity and specificity of contract requirements. This option uses an Invitation for Bid (IFB) rather than an RFP. It provides bidders with specific detailed specifications or task statements describing the contract deliverables. They tell the contractor exactly what is required and how to do it. Because there is no flexibility in the contractual task, the contract is awarded based on the low bid. This option is appropriate when the government has detailed specifications or other product baseline documentation that defines the deliverable item sufficient for production. It is generally used for simple build-to-print reprocurement.

Data Requirements

As part of the development of an IFB or RFP, the program office typically issues a letter that describes the planned procurement and asks integrated team leaders and affected functional managers to identify and justify their data requirements for that contract. The data should be directly associated with a process or task the contractor is required to perform.

The affected teams or functional offices then develop a description of each data item needed. Data Item Descriptions (DIDs), located in the Acquisition Management Systems and Data Requirements Control List (AMSDL), can be used for guidance in developing these descriptions. Descriptions should be performance based, and format should be left to the contractor as long as all pertinent data is included. The descriptions are then assembled and submitted for inclusion in the solicitation. The listing of data requirements in the contract follows an explicit format and is referred to as the CDRL.

In some cases the government will relegate the data call to the contractor. In this case it is important that the data call be managed by a government/contractor team, and any disagreements be resolved prior to formal contract change incorporating data requirements. When a SOO approach is used, the contractor should be required by section L to propose data requirements that correspond to their proposed SOW.

There is current emphasis on electronic submission of contractually required data. Electronic Data Interchange (EDI) sets the standards for compatible data communication formats.

Additional information on data management, types of data, contractual considerations, and sources of data are presented in Chapters 10 and

13. Additional information on CDRLs is provided at the end of this chapter.

Technical Data Package Controversy

Maintenance of a detailed baseline such as the "as built" description of the system, usually referred to as a Technical Data Package (TDP), can be very expensive and labor intensive. Because of this, some acquisition programs may not elect to purchase this product description. If the Government will not own the TDP the following questions must be resolved prior to solicitation issue:

- What are the pros and cons associated with the TDP owned by the contractor?

- What are the support and reprocurement impacts?

- What are the product improvement impacts?

- What are the open system impacts?

In general the government should have sufficient data rights to address life cycle concerns, such as maintenance and product upgrade. The extent to which government control of configurations and data is necessary will depend on support and reprocurement strategies. This, in turn, demands that those strategic decisions be made as early as possible in the system development to avoid purchasing data rights as a hedge against the possibility that the data will be required later in the program life cycle.

Source Selection

Source Selection determines which offeror will be the contractor, so this choice can have profound impact on program risk. The systems engineer must approach the source selection with great care because, unlike many planning decisions made early in product life cycles, the decisions made relative to source selection can generally not be easily changed once the process begins. Laws and regulations governing the fairness of the process require that changes be made very carefully—and often at the expense of considerable time and effort on the part of program office and contractor personnel. In this environment, even minor mistakes can cause distortion of proper selection.

The process starts with the development of a Source Selection Plan (SSP), that relates the organizational and management structure, the evaluation factors, and the method of analyzing the offerors' responses. The evaluation factors and their priority are transformed into information provided to the offerors in sections L and M of the RFP. The offerors' proposals are then evaluated with the procedures delineated in the SSP. These evaluations establish which offerors are conforming, guide negotiations, and are the major factor in contractor selection. The SSP is further described at the end of this chapter.

The system engineering area of responsibility includes support of SSP development by:

- Preparing the technical and technical management parts of evaluation factors,

- Organizing technical evaluation team(s), and

- Developing methods to evaluate offerors' proposals (technical and technical management).

19.3 SUMMARY COMMENTS

- Solicitation process planning includes development of a Request for Proposal, specifications, a Statement of Objective or Statement of Work, a source selection plan, and the Contract Data Requirements List.

- There are various options available to program offices as far as the guidance and constraints imposed on contractor flexibility. The government, in general, prefers that solicitations be performance-based.

- Data the contractor is required to provide the government is listed on the CDRL List.

- Source Selection is based on the evaluation criteria outlined in the SSP and reflected in Sections L and M of the RFP.

SUPPLEMENT 19-A

STATEMENT OF OBJECTIVES (SOO)

The SOO is an alternative to a government prepared SOW. A SOO provides the Government's overall objectives and the offeror's required support to achieve the contractual objectives. Offerors use the SOO as a basis for preparing a SOW which is then included as an integral part of the proposal which the government evaluates during the source selection.

Purpose

SOO expresses the basic, top-level objectives of the acquisition and is provided in the RFP in lieu of a government-written SOW. This approach gives the offerors the flexibility to develop cost effective solutions and the opportunity to propose innovative alternatives.

Approach

The government includes a brief (1- to 2-page) SOO in the RFP and requests that offerors provide a SOW in their proposal. The SOO is typically appended to section J of the RFP and does not become part of the contract. Instructions for the contractor prepared SOW would normally be included in or referenced by Section L.

SOO Development

Step 1: The RFP team develops a set of objectives compatible with the overall program direction including the following:

- User(s) operational requirements,
- Programmatic direction,
- Draft technical requirements, and
- Draft WBS and dictionary.

Step 2: Once the program objectives are defined, the SOO is constructed so that it addresses product-oriented goals and performance-oriented requirements.

SOO and Proposal Evaluations

Section L (Instructions to Offerors) of the RFP must include instructions to the offeror that require using the SOO to construct and submit a SOW. In Section M (Evaluation Criteria) the program office should include the criteria by which the proposals, including the contractor's draft SOW, will be evaluated. Because of its importance, the government's intention to evaluate the proposed SOW should be stressed in Sections L and M.

Offeror Development of the Statement of Work

The offeror should establish and define in clear, understandable terms:

- Non-specification requirements (the tasks that the contractor must do),
- What has to be delivered or provided in order for him to get paid,
- What data is necessary to support the effort, and
- Information that would show how the offerors would perform the work that could differentiate between them in proposal evaluation and contractor selection.

> **SOO Example:**
> **Joint Air-to-Surface Standoff Missile (JASSM)**
> **Statement of Objectives**
>
> The Air Force and Navy warfighters need a standoff missile that will destroy the enemies' war-sustaining capabilities with a launch standoff range outside the range of enemy area defenses. Offerors shall use the following objectives for the pre-EMD and EMD acquisition phases of the JASSM program along with other applicable portions of the RFP when preparing proposals and program plans. IMP events shall be traceable to this statement of objectives:
>
> **Pre-EMD Objectives**
>
> a. Demonstrate, at the sub-system level as a minimum, end-to-end performance of the system concept. Performance will be at the contractor-developed System Performance Specification requirements level determined during this phase without violation of any key performance parameters.
>
> b. Demonstrate the ability to deliver an affordable and producible system at or under the average unit procurement price (AUPP).
>
> c. Provide a JASSM system review including final system design, technical accomplishments, remaining technical risks and major tasks to be accomplished in EMD.
>
> **EMD Objectives**
>
> a. Demonstrate through test and/or analysis that all requirements as stated in the contractor generated System Performance Specification, derived from Operational Requirements, are met, including military utility (operational effectiveness and suitability).
>
> b. Demonstrate ability to deliver an affordable and producible system at or under the AUPP requirement.
>
> c. Demonstrate all production processes.
>
> d. Produce production representative systems for operational test and evaluation, including combined development/operational test and evaluation.

At contract award the SOW, as changed through negotiations, becomes part of the contract and the standard for measuring contractor's effectiveness.

SUPPLEMENT 19-B

STATEMENT OF WORK (SOW)

The SOW is a specific statement of the work to be performed by the contractor. It is derived from the Program WBS (System Architecture). It should contain, at a minimum, a statement of scope and intent, as well as a logical and clear definition of all tasks required. The SOW normally consists of three parts:

Section 1: Scope – Defines overall purpose of the program and to what the SOW applies.

Section 2: Applicable Documents – Lists the specifications and standards referenced in Section 3.

Section 3: Requirements – States the tasks the contractor has to perform to provide the deliverables. Tasks should track with the WBS. The SOW describes tasks the contractor has to do. The specifications describe the products.

Statement of Work Preparation and Evaluation Strategies

SOWs should be written by an integrated team of competent and experienced members. The team should:

- Review and use the appropriate WBS for the SOW framework,

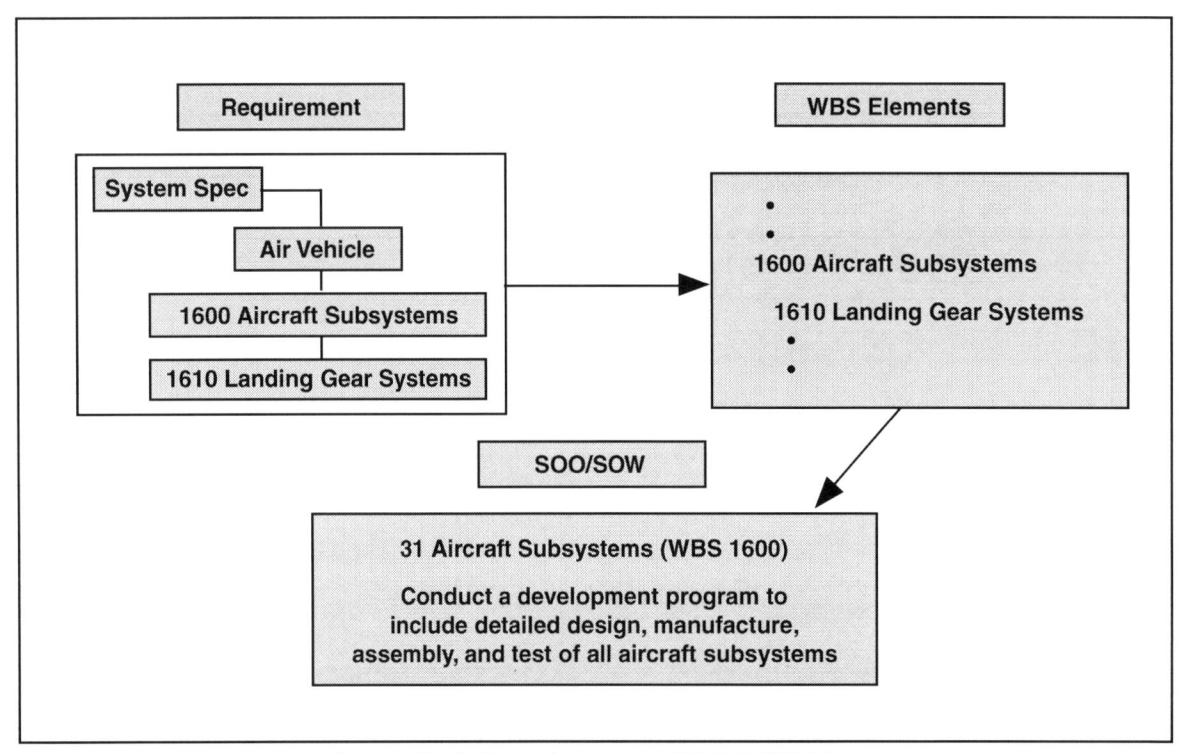

Figure 19-4. Requirement-WBS-SOW Flow

- Set SOW objectives in accordance with the Acquisition Plan and systems engineering planning,

- Develop a SOW tasking outline and check list,

- Establish schedule and deadlines, and

- Develop a comprehensive SOW from the above.

Performance-based SOW

The term *performance-based SOW* has become a common expression that relates to a SOW that tasks the contractor to perform the duties necessary to provide the required deliverables, but is not specific as to the process details. Basically, all SOWs should be performance based, however, past DoD generated SOWs have had the reputation of being overly directive. A properly developed SOW tasks the contractor without telling him how to accomplish the task.

Evaluating the SOW

The WBS facilitates a logical arrangement of the elements of the SOW and a tracing of work effort expended under each of the WBS elements. It helps integrated teams to ensure all requirements have been included, and provides a foundation for tracking program evolution and controlling the change process. As shown by Figure 19-4, the WBS serves as a link between the requirements and the SOW.

In the past, DoD usually wrote the SOW and, over time, an informal set of rules had been developed to assist in drafting them. While the government today generally does not write the SOW, but, rather, more often evaluates the contractor's proposed SOW, those same rules can assist in the government role of evaluator.

Statement of Work Rules

In section 1. *Scope:*

DO NOT:

- Include directed work statements.

- Include data requirements or deliverable products.

In section 2. *Applicable Documents:*

DO NOT:

- Include guidance documents that apply only to Government PMOs (e.g., DoD 5000 series and service regulations).

In section 3. *Requirements:*

DO NOT:

- Define work tasks in terms of data to be delivered.

- Order, describe, or discuss CDRL data (OK to reference).

- Express work tasks in data terms.

- Invoke, cite, or discuss a DID.

- Invoke handbooks, service regulations, technical orders, or any other document not specifically written in accordance with MIL-STD-961/962.

- Specify how task is to be accomplished.

- Use the SOW to amend contract specifications.

- Specify technical proposal or performance criteria or evaluation factors.

- Establish delivery schedules.

- Over specify.

In section 3. *Requirements:*

DO:

- Specify work requirements to be performed under contract.

- Set SOW objectives to reflect the acquisition plan and systems engineering planning.

- Provide a priceable set of tasks.

- Express work to be accomplished in work words.

- Use "shall" whenever a task is mandatory.

- Use "will" only to express a declaration of purpose or simple futurity.

- Use WBS as an outline.

- List tasks in chronological order.

- Limit paragraph numbering to 3rd sub-level (3.3.1.1.) – Protect Government interests.

- Allow for contractor's creative effort.

SUPPLEMENT 19-C

CONTRACT DATA REQUIREMENTS LIST

The Contract Data Requirements List (CDRL) is a list of authorized data requirements for a specific procurement that forms a part of the contract. It is comprised of a series of DD Forms 1423 (Individual CDRL forms) containing data requirements and delivery instructions. CDRLs should be linked directly to SOW tasks and managed by the program office data manager. A sample CDRL data requirement is shown in Figure 19-5.

Data requirements can also be identified in the contract via Special Contract Clauses (Federal Acquisition Clauses.) Data required by the FAR clauses are usually required and managed by the Contracting Officer.

CONTRACT DATA REQUIREMENTS LIST							
ATCH NR: 3		TO EXHIBIT:			SYSTEM/ITEM: ATF DEM/VAL PHASE		
TO CONTRACT/PR: F33657-86-C-2085			CATEGORY: X		CONTRACTOR: LOCKHEED		
1) 3100	2) SOW 3.1 3)	6) ASD/TASE	10) ONE/R	12) 60DAC	14) ASD/TASE		2/0
4) OT E62011	5) SOW 3.1	7) IT	8) D	9)	11)	13) SEE 16	ASD/TASM 2/0
^	^	^	^	^	^	^	ASD/TASL 2/0
16) BLK 4: SEE APPENDIXES TO CDRL FOR DID. THIS DID IS TAILORED AS FOLLOWS: (1) CONTRACTOR FORMAT IS ACCEPTABLE. (2) CHANGE PARAGRAPH 2a OF DID TO READ: "PROGRAM RISK ANALYSIS. THIS SECTION SHALL DESCRIBE THE PLAN AND METHODOLOGY FOR A CONTINUING ASSESSMENT OF TECHNICAL, SUPPORTABILITY, COST, AND SCHEDULE RISKS OF THE SYSTEM PROGRAM. THIS SECTION SHOULD BE CONSISTENT WITH AND NOT DUPLICATE THE SYSTEM INTEGRATION PLAN (REFERENCE DI-S-3563/T); i.e., ONE PLAN MAY REFERENCE THE OTHER." BLK 13: REVISIONS SHALL BE SUBMITTED AS REQUIRED BY CHANGE RESULTING FROM THE SYSTEMS ENGINEERING PROCESS. NOTE: SCHEDULES ASSOCIATED WITH THIS PLAN SHALL BE INTEGRATED WITH THE MASTER PROGRAM PLANNING SCHEDULE SUBMITTED ON MAGNETIC MEDIA IN ACCORDANCE WITH DI-A-3007/T.							ACO 1/0
^	^	^	^	^	^	^	15) 7/0
PREPARED BY:		DATE: 86 JUN 11		APPROVED BY:		DATE: 86 JUNE 11
DD FORM 1423 ADPE ADAPTATION SEP 81 (ASD/YYD)							

Figure 19-5. CDRL Single Data Item Requirement Example

Data Requirement Sources

Standard Data Item Descriptions (DID) define data content, preparation instructions, format, intended use, and recommended distribution of data required of the contractor for delivery. The Acquisition Management Systems and Data Requirements Control List (AMSDL) identifies acquisition management systems, source documents, and standard DIDs. With acquisition reform the use of DIDs has declined, and data item requirements now are either tailored DIDs or a set of requirements specifically written for the particular RFP in formats agreeable to the contractor and the government.

DD Form 1423 Road Map

Block 1: Data Item Number – represents the CDRL sequence number.

Block 2: Title of Data Item – same as the title entered in item 1 of the DID (DD Form 1664).

Block 4: Authority (Data Acquisition Document Number) – same as item 2 of the DID form and will include a "/t" to indicate DID has been tailored.

Block 5: Contract Reference – identifies the DID authorized in block 4 and the applicable document and paragraph numbers in the SOW from which the data flows.

Block 6: Requiring Office – activity responsible for advising the technical adequacy of the data.

Block 7: Specific Requirements – may be needed for inspection/acceptance of data.

Block 8: Approval Code – if "A," it is a critical data item requiring specific, advanced, written approval prior to distribution of the final data item.

Block 9: Distribution Statement Required:

Category A is unlimited-release to the public.

Category B is limited-release to government agencies.

Category C limits release to government agencies and their contractors.

Category D is limited-release to DoD offices and their contractors.

Category E is for release to DoD components only.

Category F is released only as directed and normally classified.

Block 12: Date of First Submission – indicates year/month/day of first submission and identifies specific event or milestone data is required.

Block 13: Date of Subsequent Submission – if data is submitted more than once, subsequent dates will be identified.

Block 14: Distribution – identify each addressee and identify the number of copies to be received by each. Use office symbols, format of data to be delivered, command initials, etc.

Block 16: Remarks – explain only tailored features of the DID, any additional information for blocks 1-15, and any resubmittal schedule or special conditions for updating data submitted for government approval.

SUPPLEMENT 19-D

THE SOURCE SELECTION PLAN

Prior to solicitation issuance, a source selection plan should be prepared by the Program Manager (PM), reviewed by the Contracting Officer, and approved by the Source Selection Authority (SSA). A Source Selection Plan (SSP) generally consists of three parts:

- The first part describes the organization, membership, and responsibilities of the source selection team,

- The second part identifies the evaluation factors, and

- The last part establishes detailed procedures for the evaluation of proposals.

Source Selection Organization

The SSA is responsible for selecting the source whose proposal is most advantageous to the government. The Source Selection Advisory Council (SSAC) provides advice to the SSA based on the Source Selection Evaluation Board's (SSEB's) findings and the collective experience of SSAC members. The SSEB generates the information the SSA needs by performing a comprehensive evaluation of each offeror's proposal. A Technical Evaluation Review Team(s) evaluates the technical portion of the proposals to support the SSEB. The process flow is shown in Figure 19-6.

The PM is responsible for developing and implementing the acquisition strategy, preparing the SSP, and obtaining SSA approval of the plan before the formal solicitation is issued to industry. The System Engineer or technical manager supports the PM's efforts. The Contracting Officer is responsible for preparation of solicitations and contracts, any communications with potential offerors or offerors, consistency of the SSP with requirements of the Federal Acquisition Regulation (FAR) and DoD FAR Supplement (DFARS), and award of the contract.

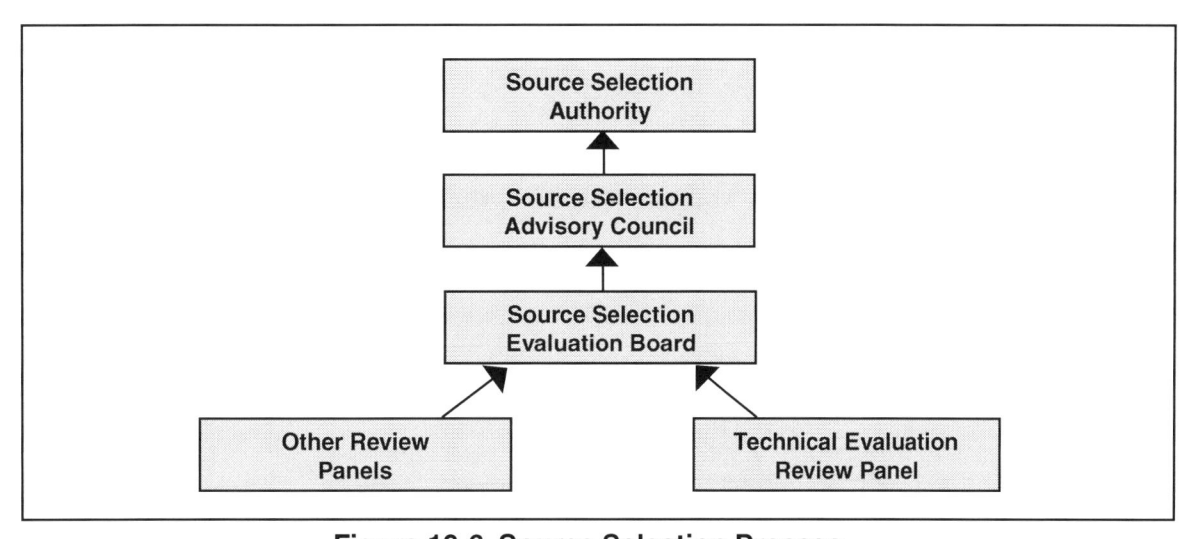

Figure 19-6. Source Selection Process

SSP Evaluation Factors

The evaluation factors are a list, in order of relative importance, of those aspects of a proposal that will be evaluated quantitatively and qualitatively to arrive at an integrated assessment as to which proposal can best meet the Government's need as described in the solicitation. Figure 19-7 shows an example of one evaluation category, life cycle cost. The purpose of the SSP evaluation is to inform offerors of the importance the Government attaches to various aspects of a proposal and to allow the government to make fair and reasoned differentiation between proposals.

In general the following guidance should be used in preparing evaluation factors:

- Limit the number of evaluation factors,

- Tailor the evaluation factors to the Government requirement (e.g., combined message of the SOO/SOW, specification, CDRL, etc.), and

- Cost is always an evaluation factor. The identification of the cost that is to be used and its relative importance in rating the proposal should be clearly identified.

Factors to Consider

There is not sufficient space here to attempt to exhaustively list all the factors that might influence the decision made in a source selection. The following are indicative of some of the key consideration, however:

- Is the supplier's proposal responsive to the government's needs as specified in the RFP?

- Is the supplier's proposal directly supportive of the system requirements specified in the system specification and SOO/SOW?

- Have the performance characteristics been adequately specified for the items proposed? Are they meaningful, *measurable*, and traceable from the system-level requirements?

- Have effectiveness factors been specified (e.g., reliability, maintainability, supportability, and availability?) Are they meaningful, *measurable*, and traceable, from the system-level requirements?

- Has the supplier addressed the requirement for test and evaluation of the proposed system element?

Rating (Points)	Evaluation Criteria – Life Cycle Cost
9-10	Offeror has included a complete Life Cycle Cost analysis that supports their proposal.
7-8	Offeror did not include a complete Life Cycle Cost analysis but has supported their design approach on the basis of Life Cycle Cost.
5-6	Offeror plans to complete a Life Cycle Cost analysis as part of the contract effort and has described the process that will be used.
3-4	Offeror plans to complete a Life Cycle Cost analysis as part of the contract effort but did not describe the process that will be used.
0-2	Life Cycle Cost was not addressed in the Offeror's proposal.

Figure 19-7. Evaluation Factors Example

- Have life cycle support requirements been identified (e.g., maintenance resource requirements, spare/repair parts, test and support equipment, personnel quantities and skills, etc?) Have these requirements been minimized to the extent possible through design?

- Does the proposed design configuration reflect growth potential or change flexibility?

- Has the supplier developed a comprehensive manufacturing and construction plan? Are key manufacturing processes identified along with their characteristics?

- Does the supplier have an adequate quality assurance and statistical process control programs?

- Does the supplier have a comprehensive planning effort (e.g., addresses program tasks, organizational structure and responsibilities, a WBS, task schedules, program monitoring and control procedures, etc.)?

- Does the supplier's proposal address all aspects of total life cycle cost?

- Does the supplier have previous experience in the design, development, and production of system elements/components which are similar in nature to the item proposed?

Proposal Evaluation

Proposal evaluation factors can be analyzed with any reasonable trade study approach. Figure 19-8 shows a common approach. In this approach each factor is rated based on the evaluation factor matrix established for each criteria, such as that shown in Figure 19-7. It is then multiplied by a weighting factor based on the perceived priority of each criteria. All the weighted evaluations are added together and the highest score wins.

Like trade studies the process should be examined for sensitivity problems; however, in the case of source selection, the check must be done with anticipated values prior to release of the RFP.

Evaluation Criteria	WT. Factor (%)	Proposal A Rating	Proposal A Score	Proposal B Rating	Proposal B Score	Proposal C Rating	Proposal C Score
A. Technical Requirements:	25						
1. Performance Characteristics	6	4	24	5	30	5	30
2. Effectiveness Factors	4	3	12	4	16	3	12
3. Design Approach	3	2	6	3	9	1	3
4. Design Documentation	4	3	12	4	16	2	8
5. Test and Evaluation Approach	2	2	4	1	2	2	4
6. Product Support Requirements	4	2	8	3	12	2	8
B. Production Capability	20						
1. Production Layout	8	5	40	6	48	6	48
2. Manufacturing Process	5	2	10	3	15	4	20
3. Quality Control Assurance	7	5	35	6	42	4	28
C. Management	20						
1. Planning (Plans/Schedules)	6	4	24	5	30	4	24
2. Organization Structure	4	4	16	4	12	4	16
3. Available Personnel Resources	5	3	15	3	20	3	15
4. Management Controls	5	3	15	3	20	4	20
D. Total Cost	25						
1. Acquisition Price	10	7	70	5	50	6	60
2. Life Cycle Cost	15	9	135	10	150	8	120
E. Additional Factors	10						
1. Prior Experience	4	4	16	3	12	3	12
2. Past Performance	6	5	30	5	30	3	18
Grand Total	100		476		516*		450

* Select Proposal B

Figure 19-8. Source Evaluation

CHAPTER 20

MANAGEMENT CONSIDERATIONS AND SUMMARY

20.1 MANAGEMENT CONSIDERATIONS

The Acquisition Reform Environment

No one involved in systems acquisition, either within the department or as a supplier, can avoid considering how to manage acquisition in the current reform environment. In many ways, rethinking the way we manage the systems engineering process is *implicit* in reforming acquisition management. Using performance specifications (instead of detailed design specifications), leaving design decisions in the hands of contractors, delaying government control of configuration baselines—all are reform measures related directly to systems engineering management. This text has already addressed and acknowledged managing the technical effort in a reform environment.

To a significant extent, the systems engineering processes—and systems engineers in general—are victims of their own successes in this environment. The systems engineering process was created and evolved to bring discipline to the business of producing very complex systems. It is intended to ensure that requirements are carefully analyzed, and that they flow down to detailed designs. The process demands that details are understood and managed. And the process has been successful. Since the 1960s manufacturers, in concert with government program offices, have produced a series of ever-increasingly capable and reliable systems using the processes described in this text. The problem is, in too many cases, we have overlaid the process with ever-increasing levels of controls, reports, and reviews. The result is that the cycle time required to produce systems has increased to unacceptable levels, even as technology life cycles have decreased precipitously. The fact is that, in too many cases, we are producing excellent systems, but systems that take too long to produce, cost too much, and are often outdated when they are finally produced. The demand for change has been sounded, and systems engineering management must respond if change is to take place. The question then becomes how should one manage to be successful in this environment? We have a process that produces good systems; how should we change the process that has served us well so that it serves us better?

At the heart of acquisition reform is this idea: we can improve our ability to provide our users with highly capable systems at reasonable cost and schedule. We can if we manage design and development in a way that takes full advantage of the expertise resident both with the government and the contractor. This translates into the government stating its needs in terms of performance outcomes desired, rather than in terms of specific design solutions required; and, likewise, in having contractors select detailed design approaches that deliver the performance demanded, and then taking responsibility for the performance actually achieved.

This approach has been implemented in DoD, and in other government agencies as well. In its earlier implementations, several cases occurred where the government managers, in an attempt to ensure that the government did not impose design solutions on contractors, chose to deliberately distance the government technical staff from contractors. This presumed that the contractor would step forward to ensure that necessary engineering disciplines and functions were covered. In more than one case, the evidence after the fact was that, as the government stepped back to a less directive role in design and development, the contractor did not take a corresponding step forward to ensure that normal engineering management disciplines were included. In several cases where problems arose, after-the-fact investigation showed important elements of the systems engineering process were either deliberately ignored or overlooked.

The problem in each case seems to have been failure to communicate expectations between the government and the contractor, compounded by a failure on the part of the government to ensure that normal engineering management disciplines were exercised. One of the more important lessons learned has been that while the systems engineering process can—and should be—tailored to the specific needs of the program, there is substantial risk ignoring elements of the process. Before one decides to skip phases, eliminate reviews, or take other actions that appear to deliver shortened schedules and less cost, one must ensure that those decisions are appropriate for the risks that characterize the program.

Arbitrary engineering management decisions yield poor technical results. One of the primary requirements inherent in systems engineering is to assess the engineering management program for its consistency with the technical realities and risks confronted, and to communicate his/her findings and recommendations to management. DoD policy is quite clear on this issue. The government is not, in most cases, expected to take the lead in the development of design solutions. That, however, does not relieve the government of its responsibilities to the taxpayers to ensure that sound technical and management processes are in place. The systems engineer must take the lead role in establishing the technical management requirements for the program and seeing that those requirements are communicated clearly to program managers and to the contractor.

Communication – Trust and Integrity

Clearly, one of the fundamental requirements for an effective systems engineer is the ability to communicate. Key to effective communication is the rudimentary understanding that communication involves two elements—a transmitter and a receiver. Even if we have a valid message and the capacity for expressing our positions in terms that enable others to understand what we are saying, true communication may not take place if the intended receiver chooses not to receive our message. What can we do, as engineering managers to help our own cause as far as ensuring that our communications are received and understood?

Much can be done to condition others to listen and give serious consideration to what one says, and, of course, the opposite is equally true—one can condition others to ignore what he/she says. It is primarily a matter of establishing credibility based on integrity and trust.

First, however, it is appropriate to discuss the systems engineer's role as a member of the management team. Systems engineering, as practiced in DoD, is fundamentally the practice of engineering management. The systems engineer is expected to integrate not only the technical disciplines in reaching recommendations, but also to integrate traditional management concerns such as cost, schedule, and policy into the technical management equation. In this role, senior levels of management expect the systems engineer to understand the policies that govern the program, and to appreciate the imperatives of cost and schedule. Furthermore, in the absence of compelling reasons to the contrary, they expect support of the policies enunciated and they expect the senior engineer to balance technical performance objectives with cost and schedule constraints.

Does this mean that the engineer should place his obligation to be a supportive team member above his ethical obligation to provide honest engineering judgment? Absolutely not! But it does mean that, if one is to gain a fair hearing for expression of reservations based on engineering judgment, one must be viewed as a member of the team. The individual who always fights the system, always objects to established policy, and, in general, refuses to try to see other points of view will eventually become isolated. When others cease listening, the communication stops and even valid points of view are lost because the intended audience is no longer receiving the message—valid or not.

In addition to being team players, engineering managers can further condition others to be receptive to their views by establishing a reputation for making reasoned judgments. A primary requirement for establishing such a reputation is that managers must have technical expertise. They must be able to make technical judgments grounded in a sound understanding of the principles that govern science and technology. Systems engineers must have the education and the experience that justifies confidence in their technical judgments. In the absence of that kind of expertise, it is unlikely that engineering managers will be able to gain the respect of those with whom they must work. And yet, systems engineers cannot be expert in all the areas that must be integrated in order to create a successful system. Consequently, systems engineers must recognize the limits of their expertise and seek advice when those limits are reached. And, of course, systems engineers must have built a reputation for integrity. They must have demonstrated a willingness to make the principled stand when that is required and to make the tough call, even when there are substantial pressures to do otherwise.

Another, perhaps small way, that engineers can improve communication with other members of their teams (especially those without an engineering background) is to have confidence in the position being articulated and to articulate the position concisely. The natural tendency of many engineers is to put forward their position on a subject along with all the facts, figures, data and required proofs that resulted in the position being taken. This sometimes results in explaining how a watch works when all that was asked was "What time is it?" Unless demonstrated otherwise, team members will generally trust the engineer's judgment and will assume that all the required rationale is in place, without having to see it. There are some times when it is appropriate to describe how the watch works, but many times communication is enhanced and time saved by providing a confident and concise answer.

When systems engineers show themselves to be strong and knowledgeable, able to operate effectively in a team environment, then communication problems are unlikely to stand in the way of effective engineering management.

20.2 ETHICAL CONSIDERATIONS

The practice of engineering exists in an environment of many competing interests. Cost and schedule pressures; changes in operational threats, requirements, technology, laws, and policies; and changes in the emphasis on tailoring policies in a common-sense way are a few examples. These competing interests are exposed on a daily basis as organizations embrace the integrated product and process development approach. The communication techniques described earlier in this chapter, and the systems engineering tools described in earlier chapters of this book, provide guidance for engineers in effectively advocating the importance of the technical aspects of the product in this environment of competing interests.

But, what do engineers do when, in their opinion, the integrated team or its leadership are not putting adequate emphasis on the technical issues? This question becomes especially difficult in the cases of product safety or when human life is at stake. There is no explicit set of rules that directs the individual in handling issues of ethical integrity. Ethics is the responsibility of everyone on the integrated team. Engineers, while clearly the advocate for the technical aspects of the intgrated solution, do not have a special role as ethical watchdogs because of their technical knowledge.

Richard T. De George in his article entitled *Ethical Responsibilities of Engineers in Large Organizations: The Pinto Case*[1] makes the following case: "The myth that ethics has no place in engineering

---

[1] *Ethical Issues in Engineering*, Johnson, Ch 15.

has been attacked, and at least in some corners of the engineering profession been put to rest. Another myth, however, is emerging to take its place—the myth of the engineer as moral hero."

This emphasis, De George believes, is misplaced. "The zeal of some preachers, however, has gone too far, piling moral responsibility upon moral responsibility on the shoulders of the engineer. Though engineers are members of a profession that holds public safety paramount, we cannot reasonably expect engineers to be willing to sacrifice their jobs each day for principle and to have a whistle ever by their sides ready to blow if their firm strays from what they perceive to be the morally right course of action."

What then is the responsibility of engineers to speak out? De George suggests as a rule of thumb that engineers and others in a large organization are morally permitted to go public with information about the safety of a product if the following conditions are met:

1. If the harm that will be done by the product to the public is serious and considerable.

2. If they make their concerns known to their superiors.

3. If, getting no satisfaction from their immediate supervisors, they exhaust the channels available within the operation, including going to the board of directors (or equivalent).

De George believes if they still get no action at this point, engineers or others are morally permitted to make their concerns public but not morally obligated to do so. To have a moral obligation to go public he adds two additional conditions to those above:

4. The person must have documented evidence that would convince a reasonable, impartial observer that his/her view of the situation is correct and the company policy wrong.

5. There must be strong evidence that making the information public will in fact prevent the threatened serious harm.

Most ethical dilemmas in engineering management can be traced to different objectives and expectations in the vertical chain of command. Higher authority knows the external pressures that impact programs and tends to focus on them. System engineers know the realities of the on-going development process and tend to focus on the internal technical process. Unless there is communication between the two, misunderstandings and late information can generate reactive decisions and potential ethical dilemmas. The challenge for system engineers is to improve communication to help unify objectives and expectations. Divisive ethical issues can be avoided where communication is respected and maintained.

20.3 SUMMARY

The material presented in this book is focused on the details of the classic systems engineering process and the role of the systems engineer as the primary practitioner where the activities included in that process are concerned. The systems engineering process described has been used successfully in both DoD and commercial product development for decades. In that sense, little new or revolutionary material has been introduced in this text. Rather, we have tried to describe this time-proven process at a level of detail that makes it logical and understandable as a tool to use to plan, design, and develop products that must meet a defined set of requirements.

In DoD, systems engineers must assume roles of engineering managers on the program or project assigned. They must understand that the role of the systems engineer is necessarily different from that normal to the narrowly specialized functional engineer, yet it is also different from the role played by the program manager. In a sense, the role of the systems engineer is a delicate one, striving to balance technical concerns with the real management pressures deriving from cost, schedule, and policy. The systems engineer is often the person in the middle; it is seldom a comfortable position. This text has been aimed at that individual.

The first two parts of the text were intended to first give the reader a comprehensive overview of systems engineering as a practice and to demonstrate the role that systems engineering plays within the DoD acquisition management process. Part 2, in particular, was intended to provide relatively detailed insights into the specific activities that make up the process. The government systems engineer may find him/herself deeply involved in some of the detailed activities that are included in the process, while less involved in others. For example, government systems engineers may find themselves very involved in requirements definition and analysis, but less directly involved in design synthesis. However, the fact that government engineers do not directly synthesize designs does not relieve them from a responsibility to understand the process and to ensure that sound practices are pursued in reaching design decisions. It is for this reason that understanding details of the process are critical.

Part 3 of the book is perhaps the heart of the text from an engineering management perspective. In Part 3, we have presented discussions on a series of topics under the general heading of Systems Analysis and Control. The engine that translates requirements into designs is defined by the requirements analysis, functional analysis and allocation, and design synthesis sequence of activities. Much of the role of the systems engineer is to evaluate progress, consider alternatives, and ensure the product remains consistent and true to the requirements upon which the design is based. The tools and techniques presented in Part 3 are the primary means by which a good engineering management effort accomplishes these tasks.

Finally, in Part 4, we presented some of the considerations beyond the implementation of a disciplined systems engineering process that the engineering manager must consider in order to be successful. Particularly in today's environment where new starts are few and resources often limited, the planning function and the issues associated with product improvement and integrated team management must move to the forefront of the systems engineer's thinking from the very early stages of work on any system.

This book has attempted to summarize the primary activities and issues associated with the conduct and management of technical activities on DoD programs and projects. It was written to supplement the material presented courses at the Defense Systems Management College. The disciplined application of the principles associated with systems engineering has been recognized as one indicator of likely success in complex programs. As always, however, the key is for the practitioner to be able to absorb these fundamental principles and then to tailor them to the specific circumstances confronted. We hope that the book will prove useful in the future challenges that readers will face as engineering managers.

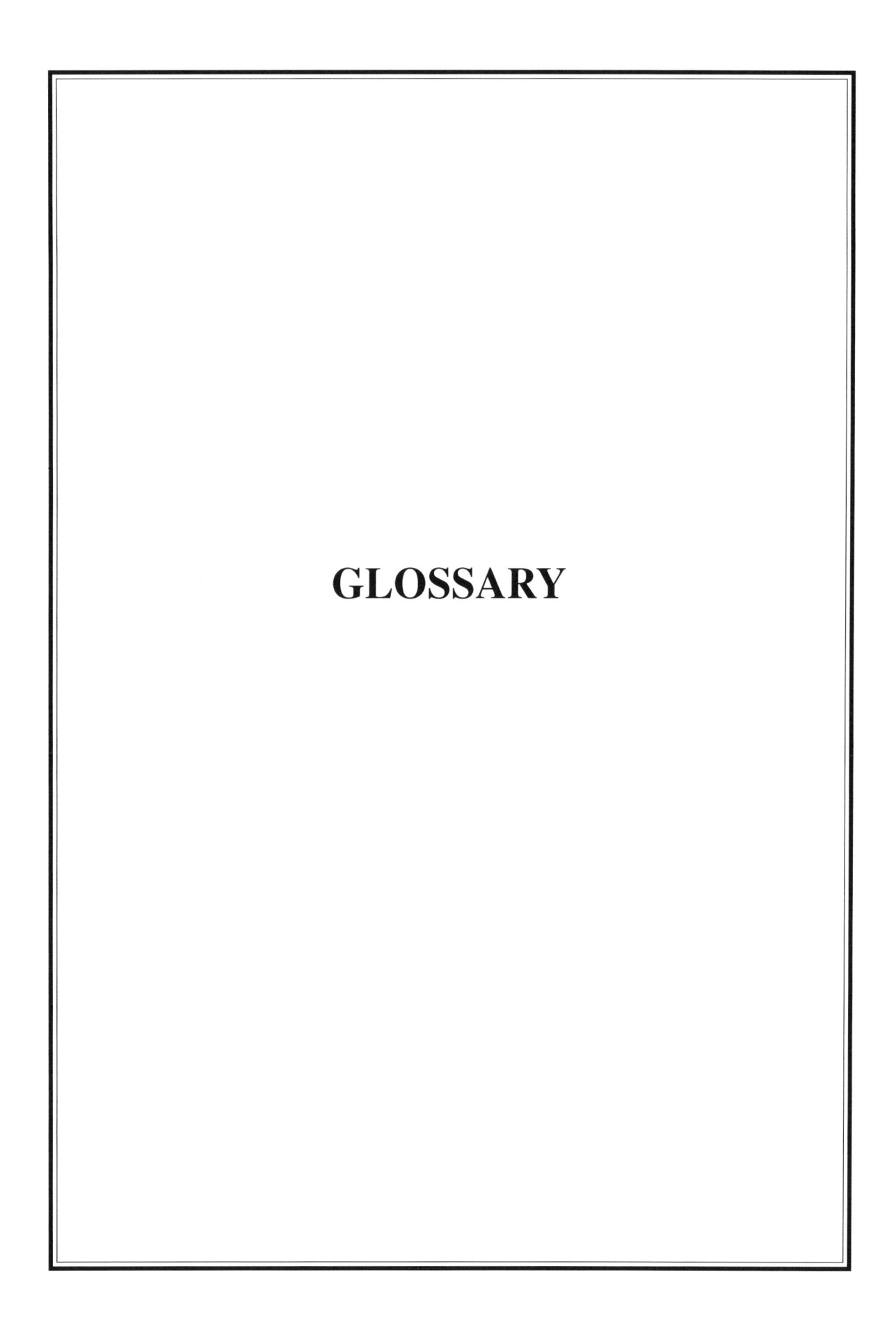

GLOSSARY

GLOSSARY

SYSTEMS ENGINEERING FUNDAMENTALS

AAAV	Advanced Amphibious Assault Vehicle
ACAT	Acquisition Category
ACR	Alternative Concept Review
AMSDL	Acquisition Management Systems Data List
ASR	Alternative Systems Review
AUPP	Average Unit Procurement Price
AWP	Awaiting Parts
BL	Baseline
BLRIP	Beyond Low Rate Initial Production
C4ISR	Command, ontrol, Communications, Computers, Intelligence, and Reconnaissance
CAD	Computer-Aided Design
CAE	Computer-Aided Engineering
CAIV	Cost As an Independent Variable
CALS	Continuous Acquisition and Life Cycle Support
CAM	Computer-Aided Manufacturing
CASE	Computer-Aided Systems Engineering
CATIA	Computer-Aided Three-Dimensional Interactive Application
CCB	Configuration Control Board
CCR	Contract Change Request
CDR	Critical Design Review
CDRL	Contract Data Requirement List
CDS	Concept Design Sheet
CE	Concept Exploration

CEO	Chief Executive Officer
CI	Configuration Item
Circular A-109	Major Systems Acquisitions
CM	Configuration Management
CM	Control Manager
COTS	Commercial Off-The-Shelf
CSCI	Computer Software Configuration Item
CWI	Continuous Wave Illumination
DAU	Defense Acquisition University
DCMC	Defense Contract Management Command
DDR	Detail Design Review
DFARS	Defense Supplement to the Federal Acquisition Regulation
DID	Data Item Description
DoD	Department of Defense
DoD 5000.2-R	Mandatory Procedures for Major Defense Acquisition Programs (MDAPs), and Major Automated Information System Acquisition Programs (MAIS)
DoDISS	DoD Index of Specifications and Standards
DSMC	Defense Systems Management College
DT	Developmental Testing
DTC	Design To Cost
DT&E	Developmental Test and Evaluation
EC	Engineering Change
ECP	Engineering Change Proposal
EDI	Electronic Data Interchange
EIA	Electronic Industries Alliance
EIA IS 632	Electronic Industries Association Interim Standard 632, on Systems Engineering
EIA IS-649	Electronic Industries Association Interim Standard 649, on Configuration Management
EOA	Early Operational Assessments

FAR	Federal Acquisition Regulation
FCA	Functional Configuration Audit
FEO	Field Engineering Order
FFBD	Functional Flow Block Diagram
FIPS	Federal Information Processing Standard
FMECA	Failure Modes, Effects, and Criticality Analysis
FOT&E	Follow-On Operational Test and Evaluation
FQR	Formal Qualification Review
GFE	Government Furnished Equipment
GFM	Government Furnished Material
ICD	Interface Control Documentation
ICWG	Interface Control Working Group
IDE	Integrated Digital Environment
IDEF	Integration Definition Function
IDEF0	Integrated Definition for Function Modeling
IDEF1x	Integration Definition for Information Modeling
IEEE	Institute of Electrical and Electronics Engineers
IEEE/EIA 12207	IEEE/EIA Standard 12207, Software Life Cycle Processes
IEEE P1220	IEEE Draft Standard 1220, Application and Management of the Systems Engineering Process
IFB	Invitation for Bid
IIPT	Integrating Integrated Product Teams
IMS	Integrated Master Schedule
IOC	Initial Operational Capability
IOT&E	Initial Operational Test and Evaluation
IPPD	Integrated Product and Process Development
IPR	In-Progress/Process Review
IPT	Integrated Product Teams

JASSM	Joint Air-to-Surface Standoff Missile
JROC	Joint Requirements Oversight Council
JTA	Joint Technical Architecture
KPPs	Key Performance Parameters
LFT&E	Live Fire Test and Evaluation
LRU	Line-Replaceable Unit
LRIP	Low Rate Initial Production
M&S	Modeling and Stimulation
MAIS	Major Automated Information System
MAISRC	Major Automated Information Systems Review Council
MBTF	Mean Time Between Failure
MDA	Milestone Decision Authority
MDAP	Major Defense Acquisition Program
MIL-HDBK-61	Military Handbook 61, on Configuration Management
MIL-HDBK-881	Military Handbook 881, on Work Breakdown Structure
MIL-STD 499A	Military Standard 499A, on Engineering Management
MIL-STD-961D	Military Standard 961D, on Standard Practice for Defense Specifications
MIL-STD 962	Military Standard 962, on Format and Content of Defense Standards
MIL-STD-973	Military Standard 973, on Configuration Management
MNS	Mission Need Statement
MOE	Measure of Effectiveness
MOP	Measure of Performance
MOS	Measure of Suitability
MRP II	Manufacturing Resource Planning II
MS	Milestone
MTTR	Mean Time To Repair
NDI	Non-Developmental Item
NIST	National Institute of Standards and Technology

NRTS	Not Repairable This Station
OA	Operational Assessment
OIPT	Overarching Integrated Product Teams
OMB	Office of Management and Budget
OPS	Operations
ORD	Operational Requirements Document
OSD	Office of the Secretary of Defense
OT&E	Operational Test and Evaluation
P3I	Preplanned Product Improvement
PAR	Production Approval Reviews
PCA	Physical Configuration Audit
PDR	Preliminary Design Review
PDRR	Program Definition and Risk Reduction
PEO	Program Executive Office
PM	Program Manager
PME	Program/Project Manager – Electronics
PMO	Program Management Office
PMT	Program Management Team
PPBS	Planning, Programming and Budgeting System
PRR	Production Readiness Review
QA	Quality Assurance
QFD	Quality Function Deployment
R&D	Research and Development
RAS	Requirements Allocation Sheets
RCS	Radar Cross Section
RDT&E	Research, Development, Test and Evaluation
RFP	Request for Proposal

S&T	Science and Technology
SBA	Simulation Based Acquisition
SBD	Schematic Block Diagram
SD&E	System Development and Demonstration
SDefR	System Definition Review (as referred to in IEEE P1220)
SDR	System Design Review
SE	Systems Engineering
Section L	Instructions to Offerors (Portion of Uniform Contract Format)
Section M	Evaluation Criteria (Portion of Uniform Contract Format)
SEDS	Systems Engineering Detail Schedule
SEMS	Systems Engineering Master Schedule
SEP	Systems Engineering Process
SFR	System Functional Review
SI	Software Item
SI&T	System Integration and Test
SOO	Statement of Objectives
SOW	Statement of Work
SPEC	Specification
SSA	Source Selection Authority
SSAC	Source Selection Advisory Council
SSEB	Source Selection Evaluation Board
SSP	Source Selection Plan
SSR	Software Specification Review
SRR	System Requirements Review
SRU	Shop-Replaceable Unit
STD	Standard
SVR	System Verification Review
S/W	Software

T&E	Test and Evaluation
TDP	Technical Data Package
TEMP	Test and Evaluation Master Plan
TLS	Timeline Analysis Sheet
TOC	Team Operating Contract
TPM	Technical Performance Measurement
TPWG	Test Planning Work Group
TRR	Test Readiness Review
VV&A	Verfication, Validation, and Accreditation
WIPT	Working-Level Integrated Product Team

Made in the USA
Monee, IL
22 August 2022